D1289292

UNLOCKING *the* MYSTERIES *of* CREATION

THE EXPLORER'S GUIDE TO THE AWESOME WORKS OF GOD

DENNIS R. PETERSEN

"One generation shall praise thy works to another,
and shall declare thy mighty acts."

Psalm 145:4

Dedication

The closer I get to the front of the line of life's journey the more profoundly I sense the urgency to invest my priorities and energy into the next generation in the line behind me. So, submitting to the Creator's sovereignty, I think it is appropriate to dedicate this work to the generation of my children, Christine, Carla, Grant and Ryan. Only the good that can be done to strengthen their faith in God's Word will have lasting value. May God use this book in the lives of multitudes of young people to give hope, inspire faith, and strengthen the resolve to serve the Savior.

Unlocking the Mysteries of Creation
The Explorer's Guide to the Awesome Works of God

Copyright © 2002 by Dennis R. Petersen

Third printing - Premier Edition

Printed in the United States of America

Art Direction by Jonathan Chong

Interior designs by Daniel Cordova and Jonathan Chong

The first English edition published in 1986 has been
reprinted over 25 times with over 100,000 copies in circulation

This new revised & expanded edition
published by
Creation Resource Publications
P.O. Box 570
El Dorado, California 95623
(530) 626-4447
On the World Wide Web visit www.creationresource.org

All scripture quotations are from the King James Version of the Holy Bible
with minimal adjustment for clarity. Other translations are noted.

Library of Congress Control Number: 2001096410   ISBN: 978-0-9672713-0-9

Acknowledgments

The forerunner of this book was a sheaf of lecture notes prepared as a team effort with my mentor, Glen S. McLean, nearly three decades ago in Canada. In God's providence, that project resulted in seminars, books and tapes that have encouraged hundreds of thousands of lives worldwide. Many others proved instrumental in making the first edition of this work a reality. God alone knows the impact their efforts have made.

This new expanded color version is also the providential result of many encouragers and helpers who deserve our humble thanks for their part in making it finally happen. Most notably, it was the patient encouragement of my associate in research, Robert Helfinstine, whose constant devotion to this project kept the vision for it strong, even through years of discouraging circumstances. Bob is a vanguard of the old school, a thorough and brilliant engineer, carefully checking every detail. His 40-year career in the aerospace industry and lifetime study of history, Scripture and science added a high level of scholarship to this project. He deserves more credit than readers can know, and I praise God for his faithful contributions and editing efforts.

Provision is unquestionably the ingredient that turns vision into reality. The Master Provider not only brought material provision through generous friends in His good time, but also prepared the life of one young artist to rescue this project at a critical stage. Without Daniel Cordova, you would not have the beauty you enjoy in this book. Dan's skill in design was made far richer by his outstanding team attitude in working tirelessly with me for over a year. Digging through thousands of images and adding many of his own, Dan built a stunning presentation on the foundations of all who helped make the first edition such a blessing. When we needed final help, the Lord brought Jonathan Chong to the project. Jonathan's technical expertise and artistic skill prepared everything for press. Don Fraser also deserves the deepest gratitude for his generous devotion to editing the final work. The blessing of both men cannot be overstated.

For the encouragement and help of so many others, I humbly recognize that this book can only be attributed to a tapestry of people woven together by Providence himself. I'm confident your attention will be drawn to the work of His hands as a result.

Dennis Petersen
El Dorado, California
Autumn A.D. 2001

"Come and see the works of God;
He is awesome in His doing toward
the children of men."

Psalm 66:5

Foreword

"Remember now thy Creator in the days of thy youth."

King Solomon of Israel, 900 BC

"Where wast thou when I laid the foundations of the earth?
Declare, if thou hast understanding."

God to Job in Arabia, 2,000 BC

"Hast thou not known? Hast thou not heard, that the everlasting God, the Lord, the Creator of the ends of the earth, fainteth not, neither is weary? There is no searching of his understanding. Even the youths shall faint and be weary, and the young men shall utterly fall; but they that wait upon the Lord shall renew their strength; they shall mount up with wings as eagles; they shall run, and not be weary; and they shall walk, and not faint."

The prophet Isaiah to Israel, 712 BC

The Rest of the Story

No matter what you discover or the amount of study you put into analyzing it, there is always more to learn about anything. That's why some brilliant students never take the risk of writing a book. Because much of what is presented here is seldom or never seen by the general public, it's natural to be skeptical. That's why the reader is constantly challenged to think for himself and check things out. Don't choose to believe or disbelieve just because of reports you heard from others, even if they were respected sources. Ask them where they got their opinions on the subject. Do your own homework. And when you discover something to be true that you thought was false, muster the courage to communicate what you've learned to those you care about. Some of the discoveries discussed here provoke disagreeing opinions, even among respected leaders. It would be nice if we could have absolute knowledge about things, but when it comes to interpreting natural discoveries there will always be challenges to our conclusions. Thus, it's wise to be appropriately tentative and open to more input as you wrestle with the observations of the physical universe. When it comes to the unseen realm, we have to be even more cautious. Deception has spoiled so many lives. But that doesn't mean we should be agnostic about everything. Don't be afraid to keep your mind open within the limits of clear absolutes. And don't forget to show patience with those who haven't climbed to your particular level of knowledge.… The climb ahead will be vastly more surprising.

Preface

"Come let us reason together," says the Lord God Jehovah through the prophet Isaiah 700 years before Christ. It's clear this will involve some thinking on our part. In our day, when political correctness and "sound bite" oversimplification have greatly "dumbed down" the general public, YOU are to be commended for picking up a book like this. Let me warn you: the information you're about to discover will not only be fascinating for you, it will also thrust you into a controversy or at least lively conversation with virtually everyone you touch.

Until recently, the popular reports of the debate over origins have blacked out and confused the scientific and historical facts about both creation and naturalism. Respectable-sounding narrators distort the truth according to the warped atheism or paganism of the intellectual elite. We all need to be reminded to consider the presuppositions of those we are hearing. Do they ignore the power and concern of the Living Creator? Do they esteem the authoritativeness of the Book of all books or dismiss it as irrelevant religion? Are they accepting other "supernatural" excuses like "aliens" and "cosmic consciousness" rather than giving credit to the miraculous power of the Redeemer of lost mankind?

This book is just an introduction to the vast subject of origins. It is intended to lay a foundation based on what the Creator has instilled in every human being: the ability to respond to truth. Is our heart, as well as our mind, open to hear from God and respond to His Word? Many of us have been so "secularized" that we are amazed when we find that the Creator has actually communicated to mankind through a miraculously conceived and preserved book: the Bible. But when we discover that truth, we begin to see everything in life differently. The deceptions become glaring. Many mysteries start to unfold. Life takes on significance, and eternity becomes more real.

Approach this material from the outset with a desire to discover the truth. Think logically with your mind and willingly with your heart. As you do I am confident the experience will be life-changing. Parents may want to walk their children through the many pages of discovery, encouraging discussion and supplemental reading along the way. Be sure to make use of the many specialty books and tapes recommended for deeper study. When the Bible becomes your foundation for understanding the world around you, everything starts to make more sense. May this be a stepping-stone to your discovery of the wonderful Lord of Life, who Himself is the One with the key to unlock the mysteries of creation.

Dennis Petersen
El Dorado, California, USA
Autumn 2001

Table of Contents

Section 3 – Unlocking the Mysteries of Original Man 124

Section 4 – Unlocking the Mysteries of Ancient Civilizations 176

Unlocking the Mysteries of the Early Earth

"You alone are the Lord; you have made heaven, the heaven of heavens, with all their host, the earth and all things on it, the seas and all that is in them, and you preserve them all. The host of heaven worships you."

Nehemiah 9:6

"For in six days
the LORD made
heaven and earth,
the sea, and all
that is in them,
and rested the
seventh day...."

Exodus 20:11

Genesis, Chapter 1

1
v. 1	Time begins
v. 1	Heaven and Earth created
v. 2	God's Spirit hovers over waters
v. 3	Light called to "be"

2
v. 7	The Expanse (or firmament) was made
v. 7	Waters separated below and above the expanse

3
v. 9, 10	Waters below the expanse gathered to become seas
v. 9, 10	Dry land appears and is called Earth
v. 12	Earth brings forth vegetation

4
v. 14, 15	Lights made in the expanse
v. 16	Sun and moon made
v. 16	Stars made
v. 17	Sun, moon and stars are placed in the expanse

5
v. 21	Great sea monsters created
v. 21	All aquatic life created
v. 21	Every winged bird created

6
v. 25	Beasts of the earth made
v. 25	Cattle made
v. 25	Everything that creeps on the dry ground made
v. 27	Man created
v. 28	The dominion over every living thing given to man
v. 29	Every plant on Earth's surface is given for food

Discovering What's True about Origins

Welcome to the great adventure of discovering what's true about origins. We all enjoy the beauty and power of the natural world, but seldom do most of us ever think about where we got our basis of understanding and interpreting it. We call that basis our "**worldview**."

Unfortunately, most of the information we hear about nature and history is from entirely secular, even atheistic sources. God and the Bible are totally left out of the picture, and often treated as legend or worse.

In his attempt to understand reality, the secular man embraces a worldview that is completely naturalistic. It leaves many major mysteries that create a lot of confusion. But is there more insight we're not hearing from ordinary sources of information?

A man named Job lived 4,000 years ago and said:

"Ask the beasts and they will teach you...

Ask the birds of heaven, let them tell you...

Speak to the earth and let it teach you...

Let the fish of the seas declare to you..." *Job 12:7-8*

What's the result of this scientific investigation? Job asks:

"Who, among all these, does not know that the hand of the Lord has done this? In His hand is the life of every living thing and the breath of all mankind." *Job 12:9*

THINK! For a man of 4,000 years ago, without a Bible or modern theology, what tremendous insight he had to the character and power of the sovereign Creator of the universe.

But what do our pagan experts who boast in their unsacred philosophy say these days? One science

magazine cover story spotlighted "The 20 Greatest Unanswered Questions of Science!"[1] Questions were raised like: "How does a single cell become a human? What happened to the dinosaurs? And **where** did life begin?" Of course, with the limitation of being a secular source, the answers to these intriguing questions remained veiled in mystery and confusion. Looking at the academic landscape around us these days, isn't it ironic that some of the most brilliant people have such a hard time discovering the Creator and Ultimate Source of all the answers? Of course the search only begins with the study of nature. Think of all the exciting mysteries awaiting our exploration in the domain of man's own past! What astounding surprises await our investigation in an attempt to discover what really happened and how such amazing accomplishments were achieved?

In these very special chapters, we'll discover surprising evidence from all over the world. First we'll examine biblical creation. Does science back it up? Genesis tells us God created "waters above the firmament" and a "very good" earth. You know what? Modern science has confirmed the simple Bible fact that our planet had a very different environment in the past that resulted in phenomenal productivity from pole to pole!

When we inquire of our earth, what **do** we learn about its beginning? Did you know that many thousands of mammoth skeletons are buried in the frozen Arctic? Well, what really happened to cause that?

What about the shifting continents? Do they somehow relate to the Bible? When you enjoy the grandeur of the mountains, what do they tell you about past events and our planet's age? The textbook charts keep insisting on millions of years to support theories of slow, gradual evolutionary change. But the facts of science support a much younger earth! Shouldn't we hear that information too?

THINK! Do you realize why the matter of Earth's age is so crucial?

The modern secular explanation for origins requires long ages of millions of years to explain the gradual naturalistic evolution of all living things. The biblical explanation, on the other hand, contains specific details that insist on a much shorter history of life on this planet. **Both cannot be true.** Thus we have a major controversy in which naturalism always questions the authority of scripture, and creationism questions the authenticity of evolutionary assumptions.

Even the heavens declare something to us, don't they? Is it evolution? We'll see as we cross-examine the claims of evolutionary thinking. We'll discover its corrupt roots, its glaring fallacies, and some of the amazing mysteries that completely unravel the very notion of the theory.

Isn't it interesting how the secular press frequently capitalizes on the fact that we all want to know where we came from! So, next in our search we'll dig up some revealing things about original man. The latest discoveries are very enlightening. Of course, we've got good reason to ask: What really is man anyway – just a bit above the apes? What do the findings of true science tell us about the amazing wonder of man's uniqueness?

Then there are those mysterious dinosaurs! We can't leave them out. Surprising clues give us insight on human contact with dinosaurs in historic times! What can we learn about the mystery of dinosaur extinction? Could global catastrophe have been responsible? What do the fossils tell us? We'll get into that later.

Finally, there is the fascinating world of ancient civilizations. Many built marvelous pyramid structures, but why? How did they do it? Just how advanced were those ancients? Why were they so capable? Can the Bible give us insight on ancient wonders like those of Babylon?

We'll explore some of the realities of the global Flood. How have violent upheavals re-arranged the look of our planet in the past? And what would a globe-rocking catastrophe do to our world if it happened today? We've seen some devastating earthquakes, hurricanes, and volcanoes in recent years, but multiply that a thousand times and think of the potential changes! With that in mind, isn't it interesting that scientists are the ones nowadays who are predicting terrifying future cosmic collisions? In times past, you would be labeled a religious doomsayer for making such predictions.

You may have noticed articles like the one that appeared in *Time* magazine.[2] The headline read "Whew! That Was Close." A huge half-mile-wide asteroid nearly knocked the world into an apocalyptic horror. They said it could happen anytime… without warning! And with the fright of that prospect, what can we do? *Time* wrote: "The most sensible thing to do about earth-grazing asteroids is try not to think about them." How comforting!

You're probably aware that the Bible specifically predicted just such physical realities long before modern scientists ever discovered the possibilities. In light of that, should anyone really be surprised to learn that the Bible is more up-to-date (as well as accurate and relevant) than the latest network news?

As we explore lots of exciting things together, we'll find that the Bible is always the key to unlocking the mysteries of creation.

How Important Is the Biblical Concept of Creation?

"In the beginning God created the heavens and the earth... Then God saw everything that He had made and indeed it was very good. So the evening and the morning were the sixth day."

Genesis 1:1, 31

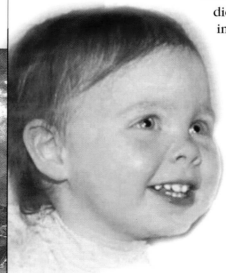

Are we to accept that these statements are open to unending speculation? Or does God intend that we are to understand that these words mean exactly what they say?

THINK! What is the most obvious question you could ask someone that could lead them to think about God? Is it not the question:

Where did I come from?

Even the most primitive tribal people ask that kind of question. Is that why the apostle Paul deals with this issue of origins in the introduction to the book of Romans in your Bible? He says there are

"men who hold the truth in unrighteousness" Romans 1:18 (KJV) *(literally "suppress the truth" as in the NASV).*

Paul says God's wrath is upon such people! Why?

Because there are facts which can be *"known about God."* (v.19). Paul insists God has shown these realities to all men! How did God do that? Well, look at verse 20, one of the most important and revealing verses in your Bible:

"From (or since) the creation of the world... the invisible things (God's attributes) even His eternal power and Godhead (or His divine nature), are clearly seen, being understood by the things that are made... so that they are without excuse."

What has been clearly seen?

God's very nature! And HOW is THAT **revealed**?

Through the things that are made!

THINK! The created physical world around us points us to God! That's why all men are without excuse for ignoring God (as it says at the end of verse 20).

The Creation Foundation

- Genesis - God's first revelation of the Bible - begins with it.

- The Gospel of John begins with it.

- Romans - the complete gospel - makes the knowledge of it the basis for judgment.

- Paul's Mars Hill discourse begins with Creator God. (Acts 17:22ff)

- Colossians begins with it.

- Hebrews establishes creation as its first basic doctrine.

But what happens when men ignore God? They still have that same probing question: "Where did I come from?" So Paul tells us what to expect to find in verse 21: *"they glorified Him not as God, but became...**vain in their imaginations**" (The NASV says "futile in their speculations")*. **THINK**! Have you ever noticed the utter futility of the reasoning of men who ignore God?

On the other hand, when you know your God is big enough to **create**, do you think that your **faith** in Him is affected? The prophet Jeremiah said:

"Ah Lord God! Behold, thou hast made the heaven and the earth by thy great power and stretched out arm, and there is nothing too hard for thee." Jeremiah 32:17

How did the infant church pray in confidence to God?

"Lord, You are God who made heaven and Earth, and the sea, and all that is in them...." Acts 4:24

Did they leave anything out? They finished by asking God to give His bondservants boldness and confidence to speak while God did signs and wonders in the name of Jesus.

"O Lord how manifold are Your works! In wisdom You have made them all! The Earth is full of Your possessions." Psalm 104:24

Creation & Our Faith

The wonderful world of nature is the creation all around us. It was designed and spoken into existence by God's very Word. He is the author of "science."

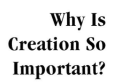

"In Him are hidden all the treasures of wisdom and knowledge" (including philosophy and science).
Colossians 2:3

The creation, so frequently used to describe God's power in both Old and New Testaments, shows us that understanding the principles of God's miraculous creation epoch is essential to our faith.

"Through faith we understand that the worlds were framed by the word of God, so that things which are seen were not made of things which do appear." Hebrews 11:3

Creation & Christ

Christ himself repeatedly acknowledged the truth of ALL of the Old Testament.

"If you believe not his (Moses') writings, how shall you believe my words?"
John 5:47

Moses penned the creation account and Jesus makes direct reference to it in Matt. 19:4 when speaking about marriage and the creation by God of male and female at the beginning.

"If they do not hear Moses and the prophets, neither will they be persuaded, though one rise from the dead." Luke 16:31

"O foolish ones and slow of heart to believe in ALL that the prophets have spoken."
Luke 24:25

Who was Israel's first prophet? (Moses)

Why Is Creation So Important?

Satan desires to undermine God's expressed account of creation. By denying creation it is easier to deny God's plan of salvation and then the truth of all God's Word.

If God is subtly removed from people's minds as their Creator, then it is simple for them to disregard Him as their Redeemer.

What Is True Science?

THINK! Isn't science supposed to deal with reality? Isn't it the study of FACTS as they really are? Pure science analyzes **The Real World**. But wait! How much of popular science is loaded with speculation and guesswork? Much of it is based on the presupposition that the creation originated by chance.

Are we really willing to program **Our Computers** with **pure knowledge**? Can such knowledge even be found? Can the truth be known? Many would challenge you on that. Remember Pilate? Before Jesus, he mocked, "What is truth?" (John 18:38).

Presuppositions: ideas assumed true without question or proof from the beginning of one's consideration of a subject. When presuppositions are based on untenable impossibilities and denial of absolute truth, the rational results are predictably nonsense.

Where is true knowledge found?

The wisest man ever (King Solomon) said:

"The fear [reverence] of the Lord is the beginning of knowledge."

Proverbs 1:7

Did you notice Solomon did not say that a bachelor of science degree is the beginning of knowledge? Yet how many well-meaning families are guided by that assertion?

How Is Your Computer Programmed?

Our minds are the most intricate computers imaginable. When we go out and look at anything in our world, we can evaluate it only on the basis of what we've already learned or heard, and accepted. Have you ever heard the phrase from computer experts:

**"Garbage In...
        Garbage Out"**

If you put misinformation in your computer, won't that confuse both your thoughts as well as your beliefs about almost everything? Do you think the devil knows that? Why do you think that old sidewinder has such a heyday when it comes to the world of education? This is why it is absolutely imperative to find out the presuppositions of your teachers before you adopt a corrupt interpretation of the knowledge you learn.

Someone once said: **"How hard, if not impossible, it is for the heart to accept what the mind rejects!"**

If we think we have to rationalize parts of God's Word because we think it's a fable, then how can we seriously believe the rest of it?

The whole Bible is built on the foundation of Genesis. Why do you think there have been such sophisticated, deceptive attacks on truths like the six-day creation, the global Flood, and many other supernatural events in Genesis.

THINK! What's the outcome of a foolish premise?

Eccl. 10:12-13 (Living Bible) states: "A fool's speech brings him to ruin. Since he begins with a foolish premise, his conclusion is sheer madness."

"DOES NOT THE EAR TRY [or TEST] WORDS" in a way similar to how our mouth tastes food? Job 12:11

Are we not failing our own sense of discernment when we fail to examine closely the foolish premises and conclusions of so-called "men of science?"

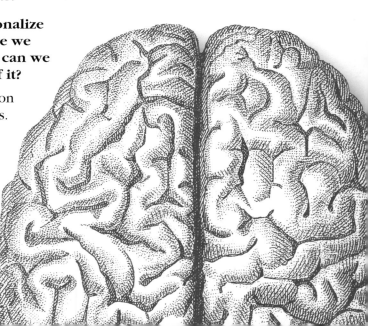

Does God Really Expect Us to Take His Word at Face Value?

By the study that God encourages (2 Timothy 2:15) you can demonstrate that the Bible is the Book of Truth. No amount of "scientific theory" can ever explain away scriptural fact!

The Word of God is:

1. Enduring: *"Forever O Lord, Your word is settled in heaven." Psalm 119:89*

"Heaven and earth shall pass away but my words shall not pass away." Matthew 24:35

2. Inerrant: *"All scripture is God-breathed." 2 Timothy 3:16 (NIV)*

"Every word of God is pure...Do not add unto His words...." Proverbs 30:5-6

"...scripture cannot be broken." John 10:35

"Let God be true, though every man be found a liar." Romans 3:4 (NAS)

3. Clear:

God's wisdom is..."all...righteousness; nothing crooked or perverse is in it..." All His words are *"plain...and right to those who find knowledge." Proverbs 8:8-9*

There's a tendency today to resist the plain straightforward absolutes of the Bible. Don't ever forget that Jesus is the One Who said: *"If they do not hear Moses and the prophets, neither will they be persuaded, though one rise from the dead" Luke 16:31.* Isn't it true that casting doubt on the clear message of the Old Testament (Torah) is crucial to the Deceiver's strategy?

What causes our lack of understanding?

You really know only what you have been taught (by observation or interpretation). If that is built on error and fallacy, then your conclusions will be distorted until some reevaluation takes place.

1. Ignorance of God's word and His power is a major problem.

Jesus said:

"Ye do err, not knowing the scriptures, nor the power of God."
                                    Matt. 22:29

2. Natural (unspiritual) thinking is another problem that affects all of us.

"..the natural man receiveth not the things of the Spirit of God: for they are foolishness unto him: neither can he know them, because they are spiritually discerned."
                                    1 Cor. 2:14

3. Attraction to fables is another human weakness.

"They shall turn away their ears from the truth, and shall be turned unto fables."   *2 Tim. 4:4*

4. Willful ignorance mustn't be forgotten either. Peter says in the last days men will be:

"willingly [or willfully] ignorant" of the creation and judgment of God. *2 Pet. 3:4-5*

5. Suppression of truth is another characteristic of our age. With all our vast knowledge today, isn't it amazing how so much truth is held back from the people?

"The wrath of God is revealed against... those... who hold the truth in unrighteousness."
                                    Romans 1:18

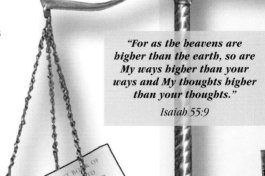

"For as the heavens are higher than the earth, so are My ways higher than your ways and My thoughts higher than your thoughts."

Isaiah 55:9

HUMAN WISDOM

What Should Our Attitude Be As We Study the Creation Around Us?

"Speak to the earth and it shall teach thee...." Job 12:8

No matter who we are or the degree of our education, it is important to examine our attitude right from the start. How do we approach our study?

1. FAITH

"Through faith we understand that the worlds were framed by the word of God...." Hebrews 11:3

Which comes first: the understanding, or the faith? If you are a Christian, do you remember how obscure the Bible seemed before you believed on Jesus Christ and were born again by faith? (Also see Luke 24:25)

2. CONFIDENCE

"Prove all things, hold fast that which is good." 1 Thessalonians 5:21

THINK! God isn't afraid of the facts! Neither should you be.

3. INSIGHT

"We wrestle not against flesh and blood, but against...spiritual wickedness in high places." Ephesians 6:12

Every conflict we encounter has an unseen dimension behind it. Winning an argument should never be our objective. We must be spiritually sensitive and prayerful if our communication is to have any truly meaningful effect.

"...the natural man cannot understand the things of the Spirit...but the spiritual man judges all things." 1 Corinthians 2:14-15

The Amplified Bible clarifies that the spiritual man appraises all things (examines, investigates, inquires into, questions, and discerns). Nothing escapes his notice. The creation by God is a very spiritual matter. It involves physical things indeed, but all of it was created by God who is spirit. It deserves being investigated.

4. OPENNESS

"The entrance of Thy words giveth light; it giveth understanding unto the simple." Psalm 119:130

The wisdom of Almighty God is available to one who humbly recognizes how imperfect his human insights are. No matter how educated we are, compared to God we just don't know very much, do we? Yet listen to God's promise for anyone with a humble attitude:

"My words are plain and clear to anyone with half a mind if it is only open." Proverbs 8:9 TLB

"Who among all these does not know that the hand of the Lord has done this? In His hand is the life of every living thing and the breath of all mankind." Job 12:9-10

What Really Happened "In the Beginning" According to the Bible?

When Did Time Begin?

"In the beginning God..."

The "beginning?" When was that? Only God himself is eternal so it is reasonable to think "the beginning" refers to the beginning of TIME, a dimension like a window in the midst of eternity. What did He do? He "created the heavens and the earth" but what does that mean? The stars weren't made until the fourth day, so what happened?

It's sort of like when you begin a construction project. What do you start with? Basic materials, of course, like lumber, nails and paint. But what raw materials did God provide first?

ATOMS? What are they? The building blocks of all physical reality, atoms are composed essentially of three things:

SPACE MATTER energy

NOTICE the words of Genesis 1:1:

"In the beginning [right at the start] GOD CREATED THE HEAVENS."

Look it up, and you'll find the original Hebrew word for heavens there is SHAMAYIM. It simply means **stretched out space!**

Think ! God even had to make the empty space in which to put everything.

Then what?

"God created...the earth."

The original word there for earth is ERETS (Hebrew for "earth") which simply means the dirt or MATTER from which everything else is made, but it was not yet as we see it today. So what was the condition of these RAW MATERIALS in the beginning? God says they were...

"without form and void."

Think ! When you take the blocks and build something, THEN you have "form." Before that the MATTER was without form.

The physical structure of matter had apparently not yet been organized and the earth was truly empty (void) of the things that were necessary for life. Also we read that *"the Spirit of God was hovering over the face of the waters."* Was this H_2O water? Since there was not yet "form" in created matter, it might be more correct to call it "fluid." Beginning with Genesis 1:3 we see God developing a **logical sequence of creative acts** to change *the "without form and void"* condition of the earth.

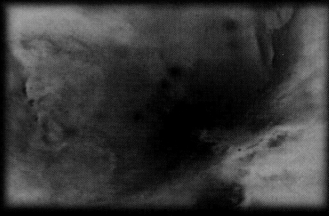

Try to imagine formlessness

And then, in that same creative instant, God said:

"Let There Be Light!"

Light includes the entire electromagnetic spectrum, not just the narrow band of color we "see." From short wave gamma rays, to long radio waves, electromagnetic radiation includes a vast range of frequencies spanning at least 75 octaves. Visible light occupies only one octave of this range. So what do we now have?

Is this not the third aspect of all physical things?

At the instant that matter was energized, basic elements took on specific form. Particles were now in motion and operating in TIME.

Think ! When physical matter ceases to be, time shall be no more. In a very simple, yet profoundly scientific way, the Genesis account of God's first creative act logically defines the basis of all physical reality:

Space, Matter & Energy

But there's more! Electromagnetic energy is at the heart of all physical matter. Atoms are held together by complex electric and magnetic forces, providing form and structure to the universe. All chemical elements took form when God said "Let there be light!" He initiated the physical laws that govern all forms of matter.

One of the mysteries of creation is the amount of energy "locked up" in the atoms. Man has been able to release some of the energy for use in atomic power plants and atom bombs, but think of the total amount of energy required to put together all the physical matter in the universe. For lack of clear evidence, some materialistic thinkers have theorized that there must be an invisible thing they call "gluons" to hold it all together. It seems a miracle of "science" that the structure of every atom in the universe doesn't fly apart.

Scientists once used the "Cloud Model" to depict the atom. Now they realize the atom is far more complex than they first imagined. The simplified "Planetary Model" was popular for most of the 20th century. The recently developed "Lucas Model" has no orbiting electrons, permitting the atom to remain stable with electric and electromagnetic forces in equilibrium.

What Does the Bible Say?

"...all things were created by Him [Jesus] and for Him; And He is before all things, and by Him all things consist [or hold together]."

Colossians 1:16,17

The Creation of Water

In Genesis 1:2, we read of God's Spirit moving (or hovering) over the face of the waters. On the beginning of the second day water covered the whole planet. God moved some of it to the space separated from the surface of the planet by what He called a firmament in Genesis 1:6-8. He then called the firmament "heavens" in the plural to apparently indicate there was more than one "heaven."

It's interesting to notice that the Hebrew word for waters is "MAYIM." The first instance of the word "mayim" is in a compounded form with the word "SHAM," meaning "there" or "in it." It is thus found as the word "SHAMAYIM" and is translated as "heavens" in Genesis 1:1. It appears there is something about the expanse of "heavens" which inherently includes water.

Scientists are learning that there is water out in the emptiness of interstellar space.[3] Could that have something to do with what are called "waters above the heavens" in Psalm 148:4?

A Parallel In Uniqueness

Water is one of the most primary of physical phenomena, yet it defies normal chemical theory. It is so unique it could not have originated by chance. With its many qualities for the support of life, it could only be the result of an incredibly complex design. The result of a strong chemical union of two gases (oxygen and hydrogen), water just happens to stay in liquid form at a narrow range of temperatures that are conducive for living creatures. When combined with carbon and certain other elements, water composes nearly all molecules of living creatures. Virtually a liquid mineral, it is a misfit to other compounds, expanding when it freezes, enabling it to float. Water is absolutely essential in liquid form for all the key systems of life. Circulation, digestion, reproduction, and respiration are all dependent on liquid water.

THINK!

"It is the Spirit Who gives life."

John 6:63 (NASV)

Remember the connection of the first Bible reference for water? It's also the first place in the Bible where God's Spirit is mentioned. This just might not be a coincidence. As water is essential for physical life, so the Spirit is essential for our spiritual life. The fact that God's Spirit interacted with water at the beginning seems to amplify the uniqueness of both in respect to their life-giving qualities.

How Important is Water to Your Life?

We all tend to take water for granted. Ask ten people if they drank the eight cups of pure water needed only to replace what is used up in normal body processes EVERY DAY. As you think about these facts, your brain cells are working and using water. Your car battery will prevent everything in your car from working if there isn't enough water in it. Likewise, dehydration leads to chronic malfunction and disease in your cells. Even the oxygen you breathe requires water to dissolve it and carry it to every cell in your body. Many health problems you develop are partly or entirely caused by lack of water. A vital key to correcting them is increasing your intake of water. Excellent research results are becoming increasingly well known. In the book, *Your Body's Many Cries for Water,* by Dr. F. Batmanghelidj, M.D., you'll see how water makes a big difference in issues like: ulcers, Alzheimer's disease, colitis, rheumatoid arthritis, headaches, allergies, stress, depression, hypertension, cholesterol, coronary health, and weight control.

Those who value wholeness in their life will do well to take water more seriously.

Without water there is NO LIFE!

Space probes have searched other planets for liquid water to see if life forms could be present. But Earth, the place where God specifically created physical life forms, is the only planet where liquid water has been found, and in abundance!

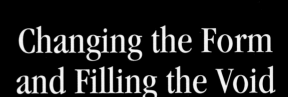

Changing the Form and Filling the Void

After the first two days of creation week, God describes for us very profound changes in the form of the primeval creation. Notice the logical preparation of the sequence with each phase providing a base and purpose for the succeeding phase.

On day 3 God changed the form of the earth by gathering the waters and letting the dry land appear. He then called the earth "good."

Next, God began to fill the earth by calling into being the grass, herbs and trees to cover the earth. The planet was no longer void, and God saw that it was good.

On the fourth day God filled the vast empty space around the earth with lights, the greater light to rule the day and the lesser light to rule the night. It is safe to presume God is referring to the sun and the moon in the context of verse 16. The stars are mentioned almost as a footnote for the fourth day of creation. And God saw that this phase too was good.

Why were the heavenly bodies made anyway?

God gives six purposes in Genesis 1 verses 14 and 15.

1. To divide the day from the night
2. For signs
3. For seasons
4. For days
5. For years
6. For lights in the firmament to give light upon the earth

Notice the last one: **"To give light on the earth."**

Think ! In God's master plan, what is the center of attention?

It's the earth!!!

The stars are there for the benefit of those on the earth! So the question we really should ask is:

Do you think a Creator God accomplishes His purposes when He says He does?

"By the word of the Lord were the heavens made and all the host of them by the breath of His mouth... for He spake, and it was done, He commanded and it stood fast" Psalm 33:6,9.

We must honestly ask ourselves if God's message here is clear or not. If He spoke and it was DONE, then there is no reason to force some explosive and destructive process into the equation. Furthermore, we have to ask if God had to wait for millions of years for the light that filled His universe from the beginning to reach viewers on the earth from distant stars.

On day 5 God created all the creatures living in the waters and all the birds of the air. God saw that this also was good.

On day 6 God created all the land animals and saw that it was good. Then He made man after His own image. After reviewing all that He had made He called the entire creation **very good.**

Think ! Why was man created last? Was it because God had created all things for the benefit of man? God had given man the earth as a complete physical and spiritual unity in fellowship with his Maker. It was a perfect place.

How Long Were
Each of the Days of Creation?

In Genesis 1, the days of the creation week are each separated by the phrase, "the evening and the morning were the first day... second day..." and so on. These are descriptive boundaries that certainly make sense to us. They would have been quite clear to Moses as well, with nothing more assumed than the straightforward meaning of the phrase. But what makes an evening and a morning? You might say "the sun." But the sun wasn't created until the fourth day along with all the other stars. Even without a sun, how would you determine the length of a day? Is it not by the rotation of the earth? If you were on another planet the length of your "day" would be determined by the rotation speed of your planet.

Of course, with only a modest knowledge of scriptural history, the implications of this would lead anyone to naturally conclude that God said "let there be light" for the first time only about 6,000 years ago. Some find this too implausible

because of the pre-conceived idea that the universe must surely be much older than that (there's that computer programming garbage surfacing again). So you might hear the assertion: "Isn't there some place in the Bible where a day is described as a thousand years long?" (as if the Bible had to make an explanation for its apparent mistake).

In the Apostle Peter's second epistle in the New Testament we find:

> *"Beloved, be not ignorant of this one thing, that one day is with the Lord, as a thousand years, and a thousand years as one day."* 2 Peter 3:8

Does Peter say a thousand years is only a day long? Of course not! And notice he's saying "with the Lord." And what's the context? Notice he's talking about judgment to come! Clearly, Peter is driving home God's patience with mankind and His over-arching presence outside the limits of time. This is not a declaration of the length of the seven days of the creation week.

So, does God's Word anywhere give us an idea of the length of the Genesis days of creation?

It's in Exodus chapter 20 when God is thundering the Ten Commandments out loud to all the Israelites at Mount Sinai. In verse eight He says, *"Remember the Sabbath day, to keep it holy."* Now why should we do that? Going on, He gives us the basis for the commandment.

> *"Six days shalt thou labor and do all thy work"* [Hebrew word for day here is "yom"]. Exodus 20:9

Why a six-day work week?

*"For in six days [same root word, "yom"] the L*ORD *made heaven and earth, the sea,* ***and all that is in them...."*** *Exodus 20:11*

Do you think God could do all that in six literal days?

Let's face it. He could have done it in six seconds if He wanted to, but He says He did it in six days! Our trouble is in not recognizing God's power. Have we created a puny concept of the Almighty Creator?

Think ! If one of those days God is talking about here really is a thousand years long, have we ever got a long work week before Saturday comes!

Before we try to fit biblical creation days into any theory requiring long ages for Earth's gradual evolution, we better take a serious look at the problems caused by such an attempt. God's revelation is clear and makes it impossible to accommodate evolutionary concepts.

According to the Bible:

1. **Earth existed before all stars!**
2. **Daylight existed before the sun.**
3. **Water covered the earth before the dry land appeared.**
4. **Complex fruit trees existed before jellyfish.**
5. **All plants existed before the sun.**
6. **Birds and whales existed before all land creatures.**
7. **Man was made before woman!**

The Genesis account simply cannot fit with the evolutionary timeline!

But what about all those stars that are millions of light years away?

Good question. The reasoning goes that in order for us to see the stars whose light supposedly took millions of years to get here (at the present speed of light) the stars themselves must have been there for millions of years. Right? Well, let's think a minute.

Why did God create the stars in the first place?

Remember they are there for the benefit of those on the earth. Did God have to wait for millions of years to achieve His purpose?

Think ! If your God is powerful enough to create the stars themselves, do you think He might be able to make their light beams instantly appear on the earth? Could the light have been created to travel at an infinitely faster speed when the creation was very good?

HOLD ON, someone may say: "But doesn't that create a misleading ... APPEARANCE OF AGE?"

Think ! How old was Adam when God created him? No matter what stage of physiological development he had at the moment he was created he would have had some appearance of age or, perhaps we should say, an **appearance of maturity!** In reality, God made Adam and Eve fully able to bear offspring and have dominion over the earth **from the beginning.**

Did God create seeds, seedlings, or fruit trees with mature fruit on them and seed in them? Read Genesis 1:11-12.

Solomon wisely observed:

"God hath made every thing beautiful in his time: also he hath set the world in their heart, so that no man can find out the work that God maketh from the beginning to the end." *Ecclesiastes 3:11*

What Are the Implications of the Origins Controversy?

Creation and natural laws

The acts of creation were governed by God and by the laws He used at creation. Those creative principles are not presently continuing for His Word declares: "And on the seventh day God ended his work which He had made; and He rested on the seventh day from all His work which He had made." (Genesis 2:2)

God finished His creation and deemed it very good before He rested. Though the laws of creation are not known, we are aware of the laws of "nature" in the created universe. God governs every natural law operating in the universe through principles that He alone established. Gravity, nuclear physics, and the dynamics of light, mass, and sound are all regulated by the **natural laws of God**. Most of them aren't understood; they are only analyzed, and then only to a limited degree.

"He upholds all things by the word of His power."
Hebrews 1:3

Do we tend to overly mystify the things of God? All of God's "**supernatural**" acts are perfectly "**natural**" to **His** nature and power! He is consistent and not capricious.

THINK! If so-called supernatural events find explanation in the framework of understood natural phenomena, does that lessen the divine miraculous wisdom behind them?

As mankind discovers, analyzes and exploits the creation, there is a tendency to make a god of knowledge and science. But for those who recognize the orderliness of God, such discovery is a gateway to the revelation of the only true Master of creation.

"...Christ, in whom are hid all the treasures of wisdom and knowledge."
Colossians 2:2-3

THINK!

"The secret things belong unto the Lord our God, but those things which are revealed belong unto us and to our children forever, that we may do all the words of this law."
Deuteronomy 29:29

Changing our Mind...

It's not easy. We are products of what we are taught, reinforced from childhood. Naturally we get defensive when encountering new concepts, especially those that contradict old concepts. But here is really an opportunity to test our spiritual and scientific maturity.

The idea of "the gradual improvement of all things" has permeated our thinking on just about every subject. Through our study we will discover that this subtle philosophy is based on dogmatic beliefs and not sound scientific integrity at all.

In the public mainstream it seems that those who do not conform to the prevailing dogma are tagged as non-conformist fools. Unfortunately this pressure by what is falsely called science has caused many Christians to re-interpret scripture in an attempt to harmonize the Bible with evolutionary theory. Thus, explanations are squeezed out of God's Word that He never intended. The word "all" is changed to "some," the definitive "did" becomes a progressive "does," and true narrative becomes allegorical poetry.

All this is done to accommodate a belief system that does not even have the qualities of a bona fide scientific theory! The whole arrangement is like a mythology of pagan cultures everywhere: you either accept the whole fantasy religiously or you're considered out of touch. But who is really out of touch?

We must objectively examine all the evidence. To that must be added all the divine revelation available to us. Then we will have a superior perspective on the real world and the One who designed and energized every delicate detail.

The Alternatives At Stake

The following is just a beginning of what could be listed of the choices one has to make when he accepts one side or the other. Add others to the list as you think of them.

Divine Creation	Spontaneous Generation
Purposeful Design	Random Accidental Order
Intricate Order	Chaotic Mistakes
Infinite God	Infinite Odds
Life from Life	Life from Non-life
God = Creator	Time = Creator
Entropy	Evolution
Catastrophe	Gradualism
True Bible	True Theories
God's Purpose	No Purpose
Absolutes Exist	Everything Is Relative
Mutations Are Harmful	Some Mutations Are Beneficial
Relatively Young Earth	Extremely Old Earth
Civilization from Start	Slow Development of Civilization
Degenerate Man	Man Getting Better
A Future Hope	Hopelessness

The Real Challenge

From the intricate universe of the microscopic cell and "invisible" atom to the unfathomable expanse of space, you look at the creation and you see order, design, and natural law.

The proof is often demanded of a Christian that an infinite God created such precision, such complexity. What's the "intellectual" alternative? Did all creation just happen on its own? Just examine the evidence and see the hallmarks of the Master Designer.

THINK! Truly, the fool has said there is no God (Psalm 14:1). And only a fool will insist, against all convincing evidence, that the only god is "eternal matter" and the only mechanism of creation is "endless ages of time."

How Good Is "Very Good"?

Have you considered how absolutely awesome and wonderful God's creation is? In the Book of Genesis in the Bible, it says that when the Creator finished His creation He declared that it was *"very good!"*

Think a moment about the nature of our temporal home, planet Earth. It's amazing, and it sure beats living on Mars.

Mars

Have you ever thought about it? Mars is only 52% farther from the sun than Earth. Temperatures there dive nearly 200 degrees below zero. The hottest day of the year is still minus 24 degrees Fahrenheit. The year is 687 days long. Nothing resembling life on Earth could thrive there.

Venus

Venus isn't much better. Due to its severely tilted axis, the daytime lasts nearly four months. The temperature there is a whopping 850 degrees Fahrenheit. Atmospheric pressure is over 1,300 pounds per square inch. Though Venus is almost the same size as Earth and is our closest neighbor, just 28% nearer the sun, it's clearly not designed for habitation.

Earth

Our Earth is so precisely placed that even a one percent variation, closer to or farther from the sun, would make life impossible. (Remember some of those hot summer days when you thought you'd bake?) And let's be thankful to a wise Creator for some of the invisible things He installed to make life more pleasant on our planet.

Magnetosphere

The magnetism of the earth causes a protective force field known as the magnetosphere. It's shaped like a huge invisible doughnut surrounding our atmosphere, thousands of miles high. Without it we'd be bombarded by harmful cosmic radiation.

Did You Know...that there is no magnetic field on the planet Venus?

Ozone

Another vital invisible component to survival on our planet is the unstable oxygen molecule called ozone. It's found largely in the highest section of our atmosphere. Ozone (or O_3 as scientists symbolize it) filters out much of the sun's ultraviolet radiation which can burn sensitive skin and cause eye damage.

Carbon Dioxide

Carbon dioxide is also invisible. Only 3/100^{ths} of one percent of our air is CO_2, yet it's absolutely essential for life. Without it plants couldn't grow, and we couldn't breathe. That's because there are special sensors in our arteries that respond to the CO_2 in the air and make you breathe. However, if the concentration rose to 10 percent, we'd all die.

Nitrogen

Oxygen is great but we ought to thank God for the 78% nitrogen in our air too. Without that high amount of nitrogen and the natural processes that work it into our soil, our planet would be desolate.

When we consider the harmful concentrations of gases like methane and ammonia on other planets like Jupiter and Saturn, we should be grateful for how carefully this Earth has been designed for life.

God's prophet Isaiah wrote:

"…the Lord that created the heavens… [is the] God that formed the earth… established it, he created it not in vain [or without purpose; He] formed it to be inhabited."   *Isaiah 45:18 (NASB)*

THINK!

• The sun is just the right size for its distance from Earth.

• The earth's size is just right and its rotation speed is too.

• The moon is just the right size for reasonable tidal variation.

• The earth's tilt is just right for the present weather cycles.

• Earth's orbital speed is just right for reasonable seasons.

• The oceans have just the right temperature variation for sea life.

• The chemistry balance is within reasonable limits for life.

• Etc., etc., etc., etc.

In short: "He formed it to be inhabited."

Job declares:

"God…doeth great things and unsearchable; marvelous things without number."

Job 5:9

What Is This "Water Above the Firmament?"

Was Earth's environment radically different at the beginning?

"And God made the firmament [or expanse] and divided the waters which were under the firmament from the waters which were above the firmament... And God called the firmament Heaven."

Genesis 1:7-8

It's easy to see that the waters under the firmament are the earth's oceans, lakes and streams, but the waters above the firmament are more challenging to identify. Could this "waters above" have been more substantial than the mere clouds we see carrying a relatively small amount of water in our atmosphere today? There are several possibilities for answering this question.

1. There has been considerable study into the concept of a water vapor canopy located high in the earth's atmosphere. One suggested reason was to provide a water reservoir for the Flood.

Could such a phenomenon really exist?

Atmospheric canopies of different sorts are found several places in our planetary neighborhood. Both Saturn and Jupiter have substantial gaseous layers.

FACT: The planet Venus is shielded by a vapor canopy so dense we cannot see through it to the surface of Venus. But this canopy is the result of the high temperature on Venus.

FACT: Saturn's moon, Titan, also has cold vapor canopies completely surrounding it.

However, no scientific observations have been discovered to support an earth canopy containing large amounts of water.

2. Suggestions of an ice-crystal canopy have been made. Isaac N. Vail introduced this basic concept in the early 1900s. Donald Cyr's recent studies of the light patterns produced by a crystal veil 50 miles above the earth, where noctilucent clouds form, have shown correlation with archeological patterns found in pottery, woven material and temple construction. Computer simulations have verified these patterns.[4] Such an ice crystal canopy would be transparent night and day except for the "sundogs" where the sun's reflections produce halos. We get the sunspot phenomenon in the winter when the air is filled with ice crystals. They are called sundogs because they follow the sun.

3. What if the waters above were not relatively close to the earth, but are still located at a great distance from Earth? Where would they be? What form would they have? When Moses, under God's direction, wrote Genesis, part of the creation account corrected false ideas that were prevalent at the time regarding the origin and characteristic of the 'waters above.' The waters above were considered to be a literal liquid ocean, and Moses did not refute this idea.[5] Also, in Psalm 148:4 the Psalmist calls for praise to the Lord from the **waters above the heavens** which must still have been in existence when he wrote those words.

Scriptural hints?

1. Genesis 2:5,6 states that God had "not sent rain" at the completion of the creation but an ideal watering system on Earth is described as a daily "mist" or fountains that rose to water the whole land. Is it significant that God's description specifically makes a

A stable worldwide paradise would require many divinely planned conditions we might have difficulty imagining today. A nearly untilted axis for the planet, denser and moister atmosphere, smaller temperature variation, and higher concentration of atmospheric carbon dioxide are some of the necessary conditions.

point of the fact that no rain occurred then, and that there is not any mention of rain until the Flood, over fifteen hundred years later?

2. Hebrews 11:7 tells that Noah was warned about "things not yet seen." Beside the Flood catastrophe itself, could the very idea of rain have also been a new thing Noah and his contemporaries experienced for the first time?

3. Genesis 9:13-14 seems to imply that the rainbow and clouds were an awesome new surprise AFTER the Flood as a sign of God's promise. If rain were a new phenomenon then a rainbow also would be an entirely new experience after the Flood.

4. Genesis 8:22 mentions the anticipation of "heat and cold" as well as "summer and winter" as God's promise as long as the earth lasts. Could extremes of seasonal climatic changes, which we commonly accept today, have been a totally new experience to Noah after the Flood?

All this leads many to believe that God originally created the planet Earth with climate conditions different from those we experience today. A relatively small vapor canopy along with a denser atmosphere with higher levels of water vapor would produce a partial greenhouse effect that would tend to stabilize the earth's temperature. Have you ever been in a big botanical garden or commercial greenhouse? It could be 30 below zero outside, but inside the greenhouse it's lush and pleasant all year around. But the greenhouse must have a source of energy to maintain its pleasant condition.

The energy source for the earth is the sun, and this energy is the controlling factor in Earth's atmospheric weather patterns, which help in distributing the energy. A controlling factor in modifying the atmospheric distribution of this energy is the tilt axis of the earth. Reducing the tilt to less than 5 degrees results in a change in airflow patterns such that warm air from the equatorial zone would flow all the way to the Polar Regions and polar air would flow to the equatorial zone. The result would be a small tropical zone, a small arctic zone and a greatly expanded temperate zone with no extremes of hot and cold.

What would be other components of a divinely designed biosphere? Besides a globally protected atmosphere, you may be correct in suggesting there would also be: increased air pressure, stronger magnetic field, slightly higher carbon dioxide and oxygen content in the atmosphere, ideal soil conditions, warmer water and other possibilities.

What could the results be?

- Longer life spans
- Minimal disease
- Rapid healing
- Greater stamina and endurance
- Larger specimens of some plants and many animals like fish and reptiles that keep growing as long as they live[6]

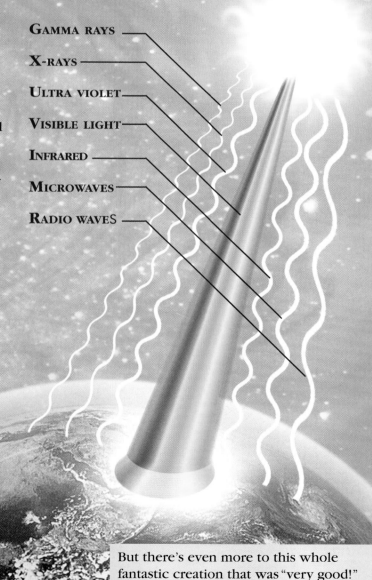

GAMMA RAYS

X-RAYS

ULTRA VIOLET

VISIBLE LIGHT

INFRARED

MICROWAVES

RADIO WAVES

But there's even more to this whole fantastic creation that was "very good!"

Does Evidence Tell of a Very Different Early Earth?

From the fossil remains of plant and animal life found encased in the layers of our planet, it is apparent that many things grew a lot larger at some time in the past than their corresponding types today.

Fossil impressions of palm fronds wouldn't be so special if they hadn't been found on northern Vancouver Island in Canada. Either the climate has changed drastically since these tropical plants grew there or some extreme catastrophic movement of these plants occurred relative to their tropical origin.

Within the Arctic Circle are two very interesting island groups. The New Siberian Islands and the Spitzbergen Islands have both been explored in the last century.

Remarkable things have been reported by explorers there. Immense frozen gravel mounds were discovered to have entombed in them entire fruit trees with the fruit still on them.[7]

Redwood forests are found buried under massive ice deposits at the South Polar region. These towering giants require a very special environment. They are typically found along the northwest coast of the

U.S.A. today, but in the past giant redwood forests flourished in many diverse parts of the world as evidenced by many coal and fossil finds.

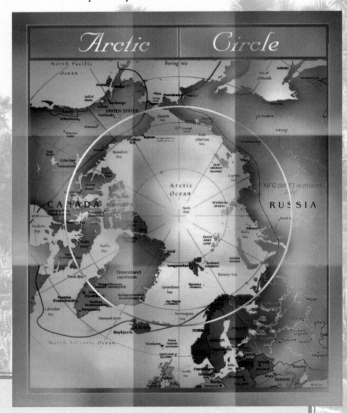

Yes indeed...The Evidence IS There!

The entire earth was very different from what we now experience.

Could the environment of the original earth have been created much more perfectly than what we witness today? Have substantial events changed all that in the history of the earth since the beginning?

Many insects thrived in proportionately larger specimens than their modern counterparts. Some cockroaches were as much as a foot long!

Dragonflies today have wings that span four inches or so. In the past, however, some of them had a wingspan of almost three feet across.

Mosses grew two or three feet in height instead of just an inch or so as they do today. Horsetail reeds today ordinarily reach heights of five or six feet. In the past similar plants grew up to fifty feet tall!

Giant animals seem to have been commonplace on Earth in the past. The hornless rhinoceros was about eighteen feet high and nearly thirty feet long!

Coiled shellfish today grow up to about eight inches across, but fossilized specimens are displayed in museums that measure over five feet across!

How Do Bones of Giant Animals Help Us Know about Earth's Original Environment?

Have you ever considered the immense wingspan of some of the giant pterosaurs (flying lizards)?

Many museums and special books about dinosaurs today create a lot of excitement over the size of some recently discovered skeletal remains. They estimate that some of the giant creatures had wingspans as broad as modern military fighter jets.

What is seldom discussed is the fact that a flying creature of any kind with a wingspan some 40 feet across would have extreme difficulty getting off the ground in the present earth's atmosphere. It couldn't run fast enough to get up enough speed to take off without serious accident. The current atmosphere is not dense enough to provide the aerodynamic lift it would need to get flying. Imagine this odd creature walking around the ground flapping its oversize wings in an effort to get airborne. If it managed to make it to the top of a cliff it could finally spread its massive glider-like appendages out and jump like a hang glider. That would do it, but think how many would crash on the rocks if they didn't have enough room to drop while they gained speed to catch the wind. No, that would never do, but there is one obvious explanation to account for the existence of these now extinct wonders of creation.

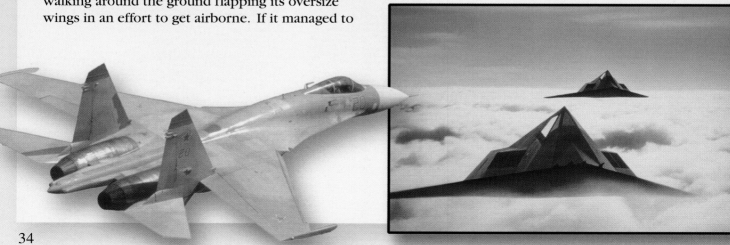

Did the earth have a higher atmospheric pressure in the past?

The gravity of the earth is capable of holding more atmosphere than we have now. Physical evidence from studies in biospheres and hyperbaric chambers indicate that our bodies heal faster in a higher atmospheric pressure. With higher air pressure, one tremendous benefit would be an increase in the rate of oxygenation of the cells. More rapid oxygen absorption means increased stamina and a quick death to viral and bacterial infections. This would answer another perennial mystery of the so-called "prehistoric" world.

How could some giant animals breathe with such relatively small chest cavities?

Lung capacity of some large dinosaurs was too small.

Dinosaur skeletons can reveal some very intriguing things about the world in which those amazing animals thrived. Could their size have required conditions on our planet that no longer exist? Some scientists have observed that when you calculate the total amount of body size that requires oxygen supply to be delivered by the animal's circulatory system, the chest space for the lungs of some of those dinosaurs just doesn't seem large enough. Even if they had an extremely rapid rate of breathing, the lungs couldn't take in enough air to provide oxygen for every extremity of the creature. So what's the answer to this mystery?

More evidence suggesting that the atmospheric pressure was once more dense.

The present atmospheric pressure measures about 14.5 pounds per square inch at sea level. A pressure of around 30 pounds per square inch would result in easier breathing and greater oxygen entry to the blood supply as it passes through the lungs. If the indications from air bubbles in ancient hardened tree sap (called amber) are correct, there might have even been as much as 25% oxygen in the air instead of the 21% measured in pure air today. These differences, along with other very good conditions in the pre-Flood world, would enable large dinosaurs to thrive and large soaring reptiles to easily lift off the ground.

What Conditions Would Make a "Very Good" Earth?

Was Earth's Magnetic Field Stronger?

The magnetic field of the earth is known to be weakening today. Scientific measurements over the years have demonstrated that the magnetic field of our planet was once much stronger than it is now.[8] What benefits would such a condition provide?

It is thought that the magnetic field of our earth is produced by a circulating current in the molten core of the planet. Like everything else in nature this built-in magnet is deteriorating with time. One of the critical benefits of Earth's magnetic field is the shielding effect provided in the earth's upper atmosphere, causing harmful cosmic rays to bounce back into space. If the magnetic field were stronger in the past, the elimination of cosmic rays would be maximized. The health of all living creatures would be improved. Most notably, an optimum magnetic field would eliminate one cause of genetic mutations leading to birth defects of many kinds.

Could World Climatic Zones Have Been Milder?

Extreme seasonal changes from summer to winter with marked variations in daily high and low temperatures are a direct result of the axial tilt of planet Earth.

Think ! Wouldn't it be nice if the mild weather of our springs and autumns lasted all year long? What would it take to do that?

The earth now has an axis tilted about 23 degrees in respect to its orbit around the sun (called the "ecliptic"). If the position were nearly vertical there would be little or no seasonal temperature change and the arctic and tropical zones would be comparatively small. The temperate zones of Earth would be greatly expanded.

What modern people seldom think about is the very real possibility and scientific evidence that the earth once did have a milder climate from pole to pole. Nothing less than a cosmic sized catastrophe would have been necessary to "kick" our planet into the axial tilt we experience now.

Atmospheric Chemistry?

With all the discussion in recent years about global warming, modern ice ages, ozone depletion and greenhouse effects, we would ask, "What is ideal?" Though trying to stop the negative results of human disturbance of the world's natural ecology may seem noble and have some positive results, there is no way to avoid the real effects of natural "entropy" acting on the global environment.

"For we know that the whole creation groaneth and travaileth in pain together until now."

Romans 8:22

The whole realm of nature is not static; neither is it improving with age. It is wearing out. There are many questions related to the environment raised by the biblical implications of a very good original earth. People lived for hundreds of years prior to a major catastrophe (the Flood) that apparently disrupted and changed many conditions into that which is less than perfect.

The Effects Of Greater Oxygen and Carbon Dioxide in the Atmosphere

We realize that oxygen is a natural byproduct of plant life on Earth. The world's massive coal beds of suddenly buried tropical forests demonstrate overwhelming evidence of ancient global flooding. The world was once filled with greenery perhaps four times more than it is today. With a generally more humid atmosphere, a 25% oxygen content would be very plausible in such a world, and very effective.

Coupled with a denser (heavier) atmosphere, increased oxygen would virtually eliminate disease. Longevity would be greatly affected. Aging could have been slowed to almost nothing. Endurance would be dramatically increased. Growth would be maximized.[9]

Carbon dioxide is a small part of the atmosphere today, but vital for the health of a flourishing plant kingdom. With only a slight percentage point of increase, the plant world would be much more successful in keeping the planet looking like a paradise.

What's the Myth Behind the Ozone Depletion Scare?

During the 1990s there was a lot of publicity about a supposedly dangerous situation in Earth's atmosphere. Many political activists claimed that modern human technology is "destroying the ozone layer" needed to filter the ultraviolet rays, thus contributing to higher rates of cancer. They said pressurized gas, leaking from refrigeration systems into the atmosphere, was causing a hole in the "ozone layer" over Antarctica.

The relatively inert compound called "Freon" is a normally safe, inexpensive, and efficient gas that is also known as a type of "CFC" (or Chlorofluorocarbon). Many politicians have been persuaded that the CFCs must be banned from production. The cost of the ban and retooling the technology for inferior substitutes for Freon is totaling billions of dollars. The call for the ban is supported by the unsubstantiated theory that the chlorine gas in CFCs escapes into the atmosphere, rises into the stratosphere, combines with ozone in the upper levels, and then breaks up the ozone shield needed to protect the earth and its inhabitants from harmful solar radiation (thus removing protection from skin cancer).

It's important to realize a common misconception. **There is no such thing as an "ozone layer" in our atmosphere!** Ozone is the name for a molecule of oxygen with an extra atom of oxygen attached to it. Thus, it is symbolized as O_3 rather than O_2. There is no "layer" of solid ozone anywhere in Earth's atmosphere, but there is a greater density of ozone molecules found in the upper 6 to 25 miles of the stratosphere. Even where it is most plentiful, there is only one molecule of ozone in every 100,000 molecules of everything else. If the entire 250-mile thick atmosphere were compressed to sea-level pressure, it would be just 5 miles thick, and ozone would only make up one-eighth of an inch of that.

The highly respected biologist and former chairman of the Atomic Energy Commission, Dr. Dixy Lee Ray wrote in her book, *Environmental Overkill*:

It is the presence of oxygen in the high atmosphere and in the stratosphere that really protects us from ultraviolet (UV) radiation…. Incoming ultraviolet radiation strikes and divides an oxygen molecule (O_2). The two separate oxygen atoms are very reactive and quickly combine with other oxygen molecules to form ozone (O_3).

Ultraviolet energy is thereby absorbed and prevented from penetrating to Earth's surface. As long as there is sufficient oxygen in the stratosphere and **as long as the sun puts out UV radiation at the right wavelength, ozone will be produced.** Several tons of ozone are produced every second.[10]

Ozone molecules are highly unstable. Various reactions including different energy wavelengths cause ozone to change back to O_2. Champions of the CFC ban claim that chlorine from CFC molecules is depleting natural ozone levels. When pressed for evidence to support the fearful forecast, none has been provided. As Dr. Ray explained **there are three major problems with the CFC-ozone depletion theory.**

First, the blocking of UV light by oxygen molecules is constantly forming ozone. There is abundant oxygen in the upper atmosphere.

Second, whatever chlorine is detected way up high is likely from evaporated seawater loaded with salt (sodium chloride), or **from the millions of tons of hydrochloric acid erupted into the air by volcanoes.**

Third, CFC presence up high is classified as "infinitesimal." Dr. Ray notes that **"CFC molecules are 4 to 8 times heavier than air."** She wrote: "We do not know how these heavier-than-air molecules cross the equatorial counter currents to accumulate at the South Pole and [supposedly] do the most ozone damage there."

THINK! If ozone is unstable and requires sunlight to be formed, why would you expect a "hole" to appear over the Antarctic at the end of the dark, cold Antarctic winter, last three to five weeks, and then disappear?

Volcanic Mount Erebus in Antarctica has released 1,000 tons of chloride daily for 20 years. It produces 50 times more chlorine annually than all the CFCs manufactured annually before CFCs were banned. The variation in the density of ozone over Antarctica appears to be associated with variations in sunspot activity, planetary waves, major storms, and "El Niño" weather patterns, as well as seasonal changes in incoming sunlight.

To complicate perceptions even more, actual documented measurements indicate a slight reduction in UV radiation on Earth in recent years, probably due to minor increases in smoke from burning which causes deflection of some sunlight. If exposure to sunlight and UV rays is insufficient, the natural process of vitamin D formation is prevented in the skin and the disease called "rickets" will occur. Another problem caused in part by too little exposure to sunlight is "brittle bones" or "osteoporosis."

There are indeed pollutants that need to be responsibly controlled. However, the ozone scare is truly misleading. The incredibly complex design of the Creator is seen in the true understanding of how ozone works in earth's atmosphere. Furthermore, when examining other deteriorating conditions in our cosmic environment like the decay of the magnetic field (see the section in this book, "When did it all begin"), how can we help but realize our total dependency on the Creator's providence and timing for our existence?"

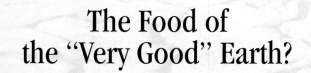

The Food of the "Very Good" Earth?

Nutrition is ultimately dependent on healthy dirt!

As rain waters and snowmelt erode the rocky mountains and hills around the world, they carry their precious dissolved cargo into the soils of the farmlands in every region. The best crops have always historically come from the lowland river deltas. Not only do those crops taste better, they provide a superior nutrition for a vigorous population. The reason is simple: optimal mineral content.

The Role of Minerals in Health

Though the modern scientific establishment is late in finally recognizing the role of natural minerals in a healthy diet, traditional people worldwide have known that crops grown on rich soils ensure the strongest and healthiest people. Animals grazed on richly mineralized pastures are unquestionably hardier and more resistant to disease than livestock struggling to forage on wasteland stubble.

Elements like copper, magnesium, calcium and about four dozen other natural raw materials are absolutely essential for healthy living cells and organs. Without those specific building blocks a body inevitably breaks down. The chronic diseases of our generation are partially, if not wholly, the result of ignoring the provision God planned for good health. We suffer needlessly by eating a diet full of deceptive food that lacks the nutrients of whole raw plants. A molecule of a living cell may have hundreds of complexly arranged atoms of carbon, hydrogen and oxygen, but the most important component in that molecule is the few atoms of **sulfur** or **iron** or other **basic minerals.** Where do the cells get those minerals?

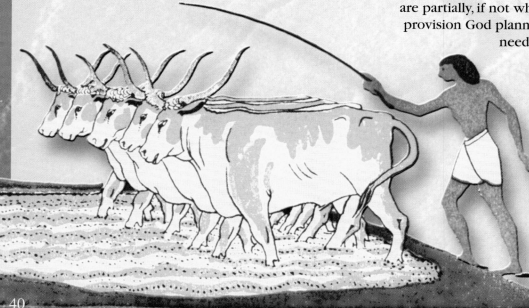

What You Eat Is What You Get

In order for minerals to get into your cells, you have to eat plants–or animals that have eaten plants–with the minerals in them.

> And God said, Behold, I have given you every herb bearing seed, which is upon the face of all the earth, and every tree, in the which is the fruit of a tree yielding seed; to you it shall be for meat [food]. And to every beast of the earth, and to every fowl of the air, and to every thing that creepeth upon the earth, wherein there is life, I have given every green herb for meat: and it was so.   *Genesis 1:29-30*

If we could digest and process dirt, then we wouldn't need to eat plants to get our minerals. But God uniquely designed plant biochemistry to draw dissolved minerals from the soil and bond them to compounds of plant cells in such a way that a human or animal can use them for cellular growth and repair. Drinking mineral water will only put excessive burden on the body's filtration system, contributing to stones and mineral accumulation in arterial tissue.

What happens when man's garden soils "wear out"?

Soil depletion of minerals is one of commercial agriculture's major problems today. Everyone knows the difference between a fresh garden-ripened tomato and a hard, pale specimen grown on over-worked commercial farm soil or in a sterile greenhouse. Though fertilizer plays an important role, many have mistakenly assumed that only a few elements like nitrogen, phosphorus and calcium are all that are needed to produce healthy crops. When rivers are allowed to occasionally flood the flood plains of the valleys, there is a continuing replenishment of dissolved minerals supplied for the crops. But when flood control dams prevent the muddy waters of flood season from covering the croplands there is a danger of reducing the nutritional potential of the produce grown there. The ancient Egyptians took pride in the strength of their soldiers who had been

fed on grains grown in the Nile River delta. The Roman army knew those crops were the best to ensure strong healthy soldiers, so control of the Egyptian lowlands was vital to the Roman Empire. However, modern man has decided to dam up the Nile, thus preventing the annual flooding and the replenishing of mineral nutrients to the farm soils.

So how did God ensure the enrichment of the soils in the "very good" earth?

Remember from Genesis 2:5-6 that there was no rain on the earth, but rather an ideal watering system of mists, or fountains, or both. God apparently caused what we might call artesian springs (full of minerals) to bubble up from deep out of the ground and water the whole surface of the planet. Thus without the storms and erosion we have now, God provided a natural way for soils to be their best and plants to be the most nutritious.

"Be sure to listen to your mother and eat your vegetables!"

What Was the Dry Land Like in the Beginning?

How was the world's landmass positioned at the start?

*Let the waters under the heaven be gathered together unto **one** place.*   Genesis 1:9

*God called the dry **land** Earth and the gathering together of the waters called he seas. [likely including lakes]*   Genesis 1:10

THINK! Is it possible God is referring here to one contiguous landmass?

We've all noticed on maps how many of the contours of the continental shores seem to fit like so many parts of a big puzzle.

In 1912, a scientist named Alfred Wegener proposed the idea of...

Continental Drift.

For 50 years his idea was ridiculed by modern scientists. But now it's approved and called the study of the world's...

Tectonic Plates.

Evolutionistic scientists, of course, start the process a supposed 200 million years ago and have the continents moving slowly apart from one another.

But could it just as well have happened suddenly through some massive upheaval of our earth?

Have you ever noticed the Bible version of this event? A descendant of Noah named Eber lived at the time of the confusion of languages at Babel.

Unto Eber were born two sons: the name of one was Peleg; for in his days was the earth divided....   Genesis 10:25

In this concise statement some awesome events are implied. The word "Peleg" literally means "to divide" or "division." The word "Earth" here speaks plainly of the geophysical earth rather than any metaphorical expression.

Most interesting is the hidden meaning of the word "divided" in this passage. It literally means to separate or "canal" by water and is associated with the word earthquake! If this is at all significant it is easy to understand why this father was so motivated to name one of his sons for this spectacular event.

Today the study of the ocean floor confirms that the landmasses **have** been ripped apart. Is it possible that the wide separation of the continents by oceans did not happen until after the division of language groups at the Tower of Babel? We're not saying this division of continents happened in moments, but instead of taking millions of years, could God have caused the separation over a few months or years?

Were volcanoes part of the earth's original environment?

When you see the typical artwork of so-called "prehistoric" scenes, you'd guess that volcanoes were a part of everyday life back then. Keep in mind what happens when a volcano erupts or explodes. Violent devastation levels the surrounding ecological systems. Death is inevitable!

Remember what God's Word says about the completed creation? It was all "very good" (Genesis 1:31). A number of Bible passages imply strongly that the curse of human and animal death did not begin until after Adam's fall in rebellion to God. Was the original earth created to be a peaceful, wonderful place for man and animals to thrive in God's blessing?

Earth's Majestic Mountains

The mountains that we know today were nonexistent in the landscape of the original earth. How can we know this?

Most of our present mountain ranges are made up of massive flood-deposited layers. Many are loaded with evidence of volcanic eruptions under water. Marine fossils abound on the tops of the highest mountains.

Though peaceful looking today, all mountains are the result of violent upheavals in the earth's crust. They show vivid evidences of a global flood. Just notice the folded and twisted layers exposed in many mountainous cliffs. These successive layers show not only a continuous deposition in a short time span, but also that the huge folds and buckled appearance had to occur while the many layers were still soft and pliable. Note the Psalmist's description of what could be the origin of the present mountains after the biblical Flood.

"Thou didst cover it with the deep as with a garment; The waters were standing above the mountains. At Thy rebuke they fled; At the sound of Thy thunder they hurried away. The mountains rose; the valleys sank down to the place which Thou didst establish for them. Thou didst set a boundary that they may not pass over; that they may not return to cover the earth."
Psalm 104:6-9

The Creation of Plants

The First Spark of Life

"And God said, Let the earth bring forth grass, and herb yielding seed, and the fruit tree yielding fruit after his kind, whose seed is in itself, upon the earth: and it was so." *Genesis 1:11*

The third day of creation

Immediately after God made the dry land appear, on the very same day, He commanded the earth to bring forth billions of plants of every description to fill the land surface of the planet. This happened even before there was the sun or any other heavenly bodies. The brilliant light of the presence of the Creator himself flooded the earth as He delicately fashioned the vast array of growing things.

The awesome little seed

To look at most seeds would hardly get anybody excited, but within that tiny factory lies the miraculous power to perpetuate life itself. Every growing thing is programmed to produce its own special seed.

Scientists can analyze and take apart the many chemical elements of the seed. But in all of its microscopic complexity, no one can build a synthetic seed. And there is no such thing as a simple one! Every seed is only produced by an amazing pattern of sexual union determined by the special construction of the parent plant.

Think how that seed can just sit there for years looking perfectly dead. Kernels of wheat stored in Egyptian tombs for thousands of years, when planted and watered, actually grew!

Never take a plant for granted again!

Ponder the specific job and intricate perfection of every part of the plant. Roots, stem, branches, leaves, flowers, and finally the fruit, each have their own special design and function.

All creatures and man need plants to survive

Only plants can take raw elements of the earth and convert them to food. Animals can't do it. This is the primary reason God made them.

Think how wonderful it is that the Creator did this with such a magnificent display of beauty as the process of growth unfolds.

Why plants?

"See, I have given you every herb that yields seed which is on the face of all the earth, and every tree whose fruit yields seed; to you it shall be for food. Also, to every beast of the earth, to every bird of the air and to everything that creeps on the earth, in which there is life, I have given every green herb for food; and it was so." *Genesis 1:29-30*

How productive?

Think how vast a crop can come from a single seed in just a few seasons. A kernel of corn yields hundreds in a few months. A dot-sized poppy seed can produce tens of thousands like itself in one summer.

Miracles of Plants

Imagine a factory that converts raw dirt plus common water, carbon dioxide and sunshine into useful and edible products! Besides foods there are textile fibers, lumber, rubber, oils, medicines and other derivatives from plants that are useful to us.

It all happens painlessly in the most efficient manufacturing system imaginable, without even a whisper of noise. There's no complicated machinery to break down. And it's been going on faithfully ever since that third day of creation.

If it weren't for plants we'd all choke to death! They exhale the oxygen we need while we dispose of the carbon dioxide they need. Think of the balance of this system so skillfully conceived by the Creator.

What a desolate place Earth would be without plants. They limit erosion, influence climate, and give shelter for birds and animals. And from the giant sequoia to the colorful fungi they make our world a continual delight of discovery and inspiration.

How productive was the original earth?

Consider the real origin of coal. Repeated sedimentary layers extending for miles with no sign of root beneath, coal seams can range from an inch thick to dozens of feet. If a worldwide, yearlong, tidal catastrophe really is responsible, then the evidence is there.

It's estimated that the world's coal reserves contain 1.4 times the carbon in the plants that would fill our Earth if it was all as lush as the tropics today![11]

When Did It All Begin?

"For in six days the LORD made heaven and earth the sea, and all that is in them..." *Exodus 20:11*

According to the genealogy lists of Genesis 5 and 11, the creation week occurred just about 6,000 years ago. Traditional history agrees with a biblical view, but the dominant evolutionary theory today demands that the earth is at least four-and-a-half or five billion years old! Have you ever noticed what a tremendous difference there is between ...

6,000 and 5,000,000,000?

WHAT'S THE DIFFERENCE?

One way to visualize the extremes of our choices is to equate one year to the thinness of one page from a typical Bible. If you were to stack up several Bibles to a height about equal with your knee, you'd have some 6,000 pages before you.

Now, how many Bibles would you have to stack up to make four-and-a-half billion pages?

The stack would reach about a hundred miles into the stratosphere.

So you're standing there between your two stacks and you are supposed to choose which one to believe in. Why is it that you are made to feel rather sheepish to admit that you lean toward the biblical stack of about 6,000 years? Or why is it that some may arrogant ridicule anyone who would dare to not agree wit their evolutionary billions?

It's so impossible to compromise these diverse positions that we are forced to realize that one these concepts is ridiculously wrong!

So which of these concepts about time is backed by historical and scientific fact?

In all honesty, we need to ask: "What is the origin of each of these ideas?" In fact we are talking about a true science.

GEOCHRONOLOGY

The science of determining the age of the earth is called "GEO-" [meaning earth] "-CHRONOLOGY (i.e. having to do with a time sequence). There are over 80 "clocks" we can examine to get an idea of how long the earth has existed.

These clocks are actual processes that keep ticking away throug the centuries. Al of them are based on the obvious reality that natural processes, occurring steadily through time, produce cumulative and often measurable results. These studies reveal maximum upper limits for the time these processes could have continued. Almost all of them point to a rather young earth. Only the least accurate and highly speculative few are misleadingly used to support billions of years. Those few are vigorously publicized to persuade the uninformed public that there is no reasonable alternative to evolutionary gradualism.

GRADUALISM

Gradualism is the evolutionary concept that present slow processes made the mountains and landforms. Dynamic, large-scale catastrophe is ruled out. Stretching present processes over millions of years supposedly accounts for it all. But if present processes can verify a relatively young earth, how will that affect our understanding and theories about the origin of our earth and, for that matter, even the universe?

All of the systems we will explore are scientifically known but generally unpublicized and unknown among even many teachers. Gradualism is the heart of evolution: slow natural processes, extrapolated for billions of years past. But what about the facts of these more-than-80 other clocks? Why is this information hidden from the general public?

JUVENILE WATER

When volcanoes erupt on the earth today, as much as 20% of the erupted material is water! This water has come from deep beneath the crust of the earth where, being under very high pressure, its temperature was extremely hot. With other gases, this water soars into the atmosphere as steam and soon precipitates in heavy rainstorms.

This water has never been on the surface of the earth before so it is called "juvenile water." Each time another volcano erupts there is more water being added to the oceans – water that was never there before.

Think ! What information can be gained from a process like this to tell us something about the beginning of things?

HOW OLD ARE THE OCEANS?

Scientists have observed volcanoes erupting at the rate of about a dozen each year. Altogether it has been estimated that their total output of juvenile water amounts to roughly one cubic mile each year.

Question! How long would it take for all the ocean water of the world to accumulate from volcanic emissions alone?

By simple math we can easily calculate backwards to find how long it would take to produce all the present water on the earth. How much water fills earth's oceans, lakes, and streams today? 340,000,000 cubic miles!

Figure it out now. At the rate of one new cubic mile of water being added each year it would take 340 million years to completely account for the origin of all Earth's surface water.[12]

What's the implication?

Based on just this one method alone, the logical conclusion is that **there were no oceans at all on the earth 340 million years ago!**

But wait!

According to the traditional evolutionary chart of Earth's history, 340 million years ago was right in the middle of the "**age of fishes**"!

Do you see the problem?

Keep in mind that the popular evolutionary idea of the origin of life assumes that the oceans were essentially full of

water at least two thousand million years ago!

What Do Processes in Nature Tell Us About Earth's Age?

"Even they will perish… and all of them will wear out like a garment."

Psalm 102:26

COMETS

When we observe comets we are witnessing a very active example of cosmic deterioration. With each elliptical circuit around the sun, every comet is a step closer to its own death. As a comet returns to the realm of Earth's orbit from its distant voyage to the neighborhood of the outer planets, a portion of its icy mass is literally blown into space by the powerful solar wind coming from the sun! On every orbit as the comet passes near the sun, some more of the comet's matter is blasted from its surface causing a spectacular "tail" that reflects sunlight and dissipates millions of miles into space. Keep in mind that the tail of the comet is blown away from the sun and is not necessarily "trailing" the path of the comet.

Knowing this, it is quite apparent that all the comets of our solar system will eventually disintegrate completely.

How long would that take?

Measuring observable comet disintegration, scientists realize that all short-period comets (with less than a 200 year orbit) would be gone in as little as 10,000 years! There are several hundred known comets in our solar system. However, seeing comets in modern times is rare when compared to the many seen in the lifetime of an average person living in Roman times. It seems likely that **comets are indeed "wearing out" pretty fast.**

Because many astronomers are not willing to admit the possibility that this process might indicate that comets, as well as the solar system itself, originated only a few thousand years ago they are forced to **dodge the issue by devising another theory.** So they suppose there must be a huge **"nest of comets"** far in the outer reaches of the solar system and, every once in a while some cosmic disturbance supposedly kicks some of these comets out of the nest.

Has anyone ever seen this nest? (They call it the Oort cloud.)

No one has ever seen a cosmic "nest" where comets occasionally spring out! Is there any evidence that comets are hatched periodically? No! What do astronomers see? If they were not actually created from the beginning for a very specific purpose (perhaps as a heavenly sign as indicated in Genesis 1:14) comets might be the result of a violent planetary explosion in the last few thousand years. They are dying, but no new ones are appearing. However, such explanations as comet "nests" must be invoked if one feels constrained to avoid the glaringly obvious conclusion that…

the planets and the comets just haven't been there very long.

OIL DEPOSITS PRESSURE

What happens when oil well drillers hit a pocket of oil deep in the earth? It's possible for a "gusher" to spray crude oil into the air for days or even weeks because of the tremendous pressure trapped below the earth's surface in those sedimentary rocks.

THINK! Even the densest sedimentary rocks have some degree of porosity. With time, what would happen to the oil pressure?

Naturally, it would dissipate into the surrounding rock formations. And the time that would take is measured in thousands of years, not millions! Tremendous pressures are not unusual in very deep wells. If those oil deposits had been there for more than 5,000 years, in some

cases there would be no pressure left! It's sort of like a slow leak in an old tire. Eventually the pressure dissipates and the tire goes flat.

The only objective explanation is that these oil deposits were suddenly and catastrophically encased in these flood-produced layers just a few thousand years ago.[13]

Where's All the Dust?

INTERPLANETARY DUST

Did you realize that cosmic dust from outer space is falling right now all over the surface of our planet? It amounts to millions of tons per year worldwide.[14] That's because there is a sizeable amount of dust floating around in the space between the planets of our solar system.

Where did this dust originate?

Because this fine cosmic dust is high in nickel content it may be supposed that it is part of the debris left from the same explosion of a planet between Mars and Jupiter that created the asteroid belt. Of course we don't find much of this dust on the surface of the earth because it keeps washing into the sea. So how much evidence of nickel is there in the sea?

Are there millions of years worth of this dust dissolved in the sea?

As a matter of fact, there is about as much nickel dissolved in seawater as would be produced by the rivers eroding the continents for only 18,000 years at the present rate of erosion.[15] What's more, there is no indication that there are enough of minerals like nickel in the seawater to cause them to precipitate into deposits gathering on the sea floor. So the evidence we can actually examine would make us suspect that the process must not have been going on very long, certainly not for millions of years. So what do evolutionists suggest to keep supporting their millions of years?

We're told the ocean floors are slowly "sub-ducting" under the continents. So the dust that is supposed to have been precipitating out of the seawater for millions of years has been "swept under the carpet" so to speak. If that is true, then of course there is no way to ever find the missing cosmic dust. However, that brings up another controversy. Where else should we expect to see cosmic dust accumulating?

How old is the moon?

The moon is an ideal laboratory for study because it's supposed to be as old as the earth, but there's no water and no hydro-erosion there. Since that fine powdery cosmic dust has been gathering everywhere on the moon's surface for supposedly close to five billion years, can you see why the space agency was concerned?

THINK! Why did the first lunar landing module have big wide saucers at the base of each leg?[16]

NASA's Biggest Worry?

NASA scientists were very concerned about their landing mission to the moon. Some estimated there would be a one-hundred-foot thick layer of dry cosmic powder all over the moon. Even conservative calculations expected a layer 54 feet thick. With fears of losing the astronauts in a poof of deep powder, NASA did everything possible to make a very gentle landing. That's why the huge saucer "pods." They even built sensing rods extending about three feet straight down under those saucer-like landing pads to measure the subsurface temperature of the lunar dust. But what did astronauts actually find when they got there? About one eighth to three inches of dust was all that was there! How long would it take to gather that much dust? At historically measured rates research suggests it would only take...

About 8,000 years MAX!

The direct implication is fairly simple. However, current scientific thought does complicate the matter. The reason the dust was an issue originally is because of the belief that dust was accumulating for millions of years, even billions. However, if the dust is a result of a more recent planetary explosion, then there wouldn't be as much dust as was expected. Furthermore, the bombardment of the lunar surface by millions of small and large meteors over the years is expected to have caused fusing and solidification of the dust that had previously landed.

... one giant step for mankind ...

When the first astronaut stepped on the moon, leaving this footprint, he expressed surprise that so little dust was there compared to what was expected.

So, can we know the age of our moon? The dust argument is inconclusive, but there certainly is no substantial evidence to insist that the moon is millions of years old.

What Does the Land Show Us About the Beginning?

EROSION

Think of the continuing erosion that tears down the earth's continents, moving millions of tons of earth into the oceans each year.

The indications of continental geography show that past rates of erosion were much greater than today's rate. But even so, if millions of years truly occurred, at the present rate of erosion, **there should have accumulated at least 30 times more sediment in the ocean than there is actually found.** Of course to believe that, you have to assume that the ocean has been there for at least a billion years.

Even more surprising is the discovery that **at the present rate of erosion all the continents on Earth would be worn down to sea level after just 14 million years.** But there's no evidence such drastic erosion even happened once.

Look at the mountain ranges. In the Rocky Mountains of North America there are billions of fossilized animals in the rock layers which are claimed to be 100s of millions of years old. And they are thousands of feet above the ocean level.

THINK! How many times would these high mountains have worn down to sea level in the last 100 million years? How did those fossils manage to stay undisturbed all that time while the mountains could have eroded down to sea level six or seven times?

The mountains and valleys on Earth appear to be rather recently formed. Their sharp angular appearance testifies to their youthfulness.[17]

TOPSOIL

One writer observed that *"The soil which sustains life lies in a thin layer of an average depth of seven or eight inches over the face of the land; the earth beneath it is as dead and sterile as the moon. That thin film is all that stands between man and extinction."*

THINK! How long does it take for topsoil to accumulate?

Scientists estimate that the combination of plant growth, bacterial decay, and erosion produces six inches of topsoil in 5,000 to 20,000 years.[18]

If the earth has been going on about the same as it is today for millions of years, one wonders why there isn't a whole lot more topsoil than there really is. There should be hundreds of feet of it. **Maybe this is just another sign the earth hasn't been here long.**

Rivers Tell a Story Too

The watershed of the Mississippi

Like all the rivers of the world, the watershed of the Mississippi River transports a tremendous amount of silt downstream to the ocean every year. Where does it end up? In this case it keeps enlarging the delta that extends into the Gulf of Mexico. Every year the delta deposit is enlarging by a known amount.

A century ago, even the evolutionists were aware that millions of years of erosion would seem to make the delta extend all the way to Africa! Why? **At the present rate the entire Mississippi River delta would accumulate in only 5,000 years!**[19] Do you realize the direct implications of that? What was there before that? Well, the river couldn't have been there could it? But science acknowledges that the river has been even bigger in the past. How could this be? Unless of course the North American continent, and all the other continents for that matter, just haven't been in their present positions any longer than that.

THINK! What did happen to this planet less than 5,000 years ago that could have rearranged all the continents?

NIAGARA FALLS

This famous waterfall is a magnificent example of a geo-clock that reveals a very young earth.

Because the rim of the falls was wearing back at a known rate every year (before man stabilized it in recent times), geologists recognize that it **has only taken about 5,000 years to erode** from its original precipice.[20]

CORAL REEFS

The buildup of the calcium carbonate remains of marine creatures in the warm oceans of our world could be accounted for entirely in the few thousand years since the worldwide Flood.[21]

STALACTITE GROWTH IN CAVES

If you've toured in a limestone cave you were likely told that the formations of "dripstone" developed very slowly over a period of more than one hundred thousand years. What is the evidence?

Under the Lincoln Memorial in Washington D.C., stalactites had grown to five feet in less than 50 years. Other evidence shows that cave formations could be easily accounted for in tens of thousands of years at the most.[22]

IGNEOUS CRUSTAL BUILD-UP

With today's average of a dozen volcanic eruptions a year, there is a steady addition of new igneous rock. There is more volcanic activity on the ocean floor that cannot be measured. The observation of our geography shows there have been times in the past of much more intense volcanism.[23]

Conservatively looking at the statistics, **the entire crust of the earth** could have developed without any other process besides volcanism **in only 500 million years.** Could the land have been missing in the Cambrian period? Or is it possible that the dry land was created a relatively short time ago?

Can Living Things Verify Earth's Age?

What is the oldest living thing on Earth?

Most of us have heard about the antiquity of the giant redwood trees in California. These massive living towers have grown to more than 300 feet tall. Some of them were already 2,000 years old when Jesus Christ ministered in ancient Galilee.

But there are other plants alive today that date back further yet.

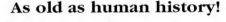

As old as human history!

The twisted and weather-beaten bristlecone pine trees cling stubbornly to life in one of the most hostile environments on Earth where life can exist. In the White Mountains bordering California and Nevada, high in the arid desert, these rare and rugged trees have been growing for about 5,000 years! Their annual growth rings have been studied to give a reasonably accurate idea of their beginnings.[24]

The bristlecone trees are so hardy it is fair to say that they will likely go on living for additional thousands of years, unless some catastrophe destroys them.

THINK! Why don't we find a grove of trees somewhere in the world dating back to 8,000 years, or 10,000 or 15,000 years?

If trees like this have lived 5,000 years they could have certainly lived longer. It's almost as though all these trees were planted on a virgin earth just 5,000 years ago!

The Bible gives a clear historical record pointing to the global Flood about 5,000 years ago. With the entire earth desolated, it stands to reason why the trees of greatest longevity would date back only to that time but no further.

What do population studies show?

One of the most revealing "clocks" deals with the growth of Earth's population down through time. Dr. Henry Morris brilliantly addresses the subject in his book, *Scientific Creationism,* which is highly recommended. Consider the following facts Dr. Morris has gathered.

THINK! If man has been on Earth for a million years, as evolutionists believe, why is population explosion only recently becoming a problem?

Today — Worldwide

- Families average 3.6 children
- Population grows 2% annually

FACT: The present population would have developed from a single family in just 4,000 years if the growth rate were reduced to only ½ per cent per year or about an average of only 2 ½ children per family per generation.

That's a fourth the present rate of growth. It would easily allow for long periods of no growth due to famines, wars, epidemics and natural disasters.

What does the evolutionary framework have to offer?

Did You Know? Evolutionists insist humans have been populating the earth for a million years! Do you have any idea how many people there would be on earth in a million years, even if the population growth rate was reduced to one fourth the present annual average (or a scanty ½ per cent a year)? With a supposed million-year history of man there would have been an incredible 25,000 generations (at 40 years each).

Even more incredible is the fact that the final total of people amounts to only the present population of about six billion.

Does this fit the statistical facts?

How big would the population be now if it increased only ½ per cent per year for a million years? In other words we would have to insist that there be an average of only 2.5 children per family for 25,000 consecutive generations.

At that rate the present population would be represented by the number 10 with 2,100 zeros after it!

Obviously, that is impossible. If you filled the entire universe with tiny electrons, you could only pack in the number 10 with 130 zeros after it!

Even if a million years of man living on the earth could somehow produce only the present population, **how many people would have lived and died in all that time?**

It would have been at least 3,000 billion people!

THINK! Where are all the bones and artifacts?

That's at least a couple of dozen graves for every acre on earth! But ancient bones and artifacts are extremely rare.[25] It seems that the facts line up directly with the biblical model for man's origin.

A basic study of biblical history reveals that the Genesis flood happened less than 5,000 years ago, when every human being outside the ark is supposed to have died.

The most accurate verifiable studies of human history place the origins of cultures like Egypt and China back less than 5,000 years ago.

The growth rate of human population fits naturally into a creation and Flood framework. The evolution model has absolutely no reasonable solution to offer.

What Other Systems Declare the Earth Is Young?

The Magnetic Field

The phenomenon that makes compasses point north and south reveals Earth's maximum age. Like everything else in the universe, the principle of progressive deterioration (second law of thermodynamics) is operating here too. Even magnets wear out, don't they?

The earth is not a permanent magnet. It is a complex electromagnet. Earth's interior is too hot for the metallic core (mostly iron) to be a permanent magnet. If you study magnetic declinations for different points around the world you'll see variations for magnetic north. These numbers continually change. Airport runway designations are changed regularly to identify the current magnetic heading.

The brilliant physicist, Thomas Barnes, shows that Earth's magnetic field has decayed regularly since it was first measured in 1835. Every 1,400 years the field strength decreases by half. Careful measurements worldwide reveal the strength of the field must have been much greater in the past. That would be fine back as far as 6,000 years, but if you project back 10,000 years ago we run into serious difficulties. The earth's magnetic field would then have been equal to that of a magnetic star. According to Dr. Barnes, we can only conclude that life on earth would not have been possible more than 10,000 years ago.[26]

The weakening magnet of our planet will eventually cause the disappearance of the magnetosphere and the accompanying Van Allen radiation belts. These are invisible but vitally necessary to support life. There are no such structures around the planet Venus because Venus has no magnetic field.

How much longer can planet Earth support life?

Many things could change catastrophically to make life impossible on this delicate planet. As Earth's magnet weakens, the protective invisible "blanket" will break down and allow increasing amounts of genetically harmful cosmic rays through our atmosphere. Some scientists worry that the situation is already taking its toll and that several thousand years at most is all the time left for life to survive here.

What about Magnetic Reversals?

Many reports say the field has reversed about every million years over the past several hundred million years. This is misleading. It is a conclusion, not an observation. It does not fit the young earth model. Evidence from magnetized rocks in deep fossil layers on the earth's continents shows that the earth's magnetic field reversed its direction on a global scale about a hundred times while these strata were being deposited. Contrary to the long age views of millions of years for the deposit of these strata, there is strong evidence that the reversals took only a few days apiece.

Such evidence supports a theory by Dr. Russ Humphreys that turbulence in the earth's fluid core would cause many rapid reversals during the single year of the Genesis Flood. Such turbulence in the core would be a normal consequence of the Catastrophic Plate Tectonics theory of the Flood proposed by Dr. John Baumgardner and other creationist scientists.[27] If veteran earth scientist Bill Overn of St. Paul, Minnesota is correct, the turbulence and observed reversals would have naturally resulted from a sudden change in the tilt axis of the earth. Such a change has been documented by George F. Dodwell and reported by Barry Setterfield at the 1983 National Creation Conference in Minneapolis. Results indicate that the change happened at the time of the flood.[28]

Dissolved Minerals in the Ocean

As worldwide erosion processes continue, Earth's minerals are dissolved and carried by rivers into the oceans. The concentrations of these dissolved minerals are increasing slightly every day.

Since we can measure the amounts of these minerals in the river waters and in the oceans, we can also calculate how long it would take, at present rates, for these elements to reach their present concentrations in the oceans.

Of all the minerals and compounds found in seawater, none of their present concentrations require the assumed evolutionary age for the ocean! Yet the evidence does not suggest the elements have precipitated out of solution. The only reasonable conclusion is that the oceans are relatively young.[29]

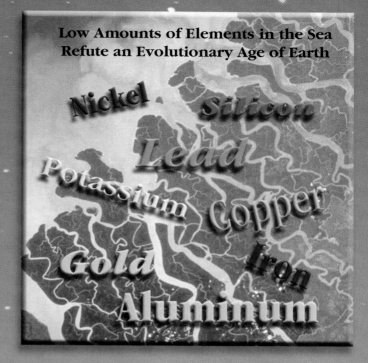

Low Amounts of Elements in the Sea Refute an Evolutionary Age of Earth

Nickel Silicon Lead Potassium Copper Gold Iron Aluminum

Atmospheric Helium

The light gas, helium, used to fill balloons, is steadily gathering in the outer reaches of our atmosphere. The total amount can be measured. One source of helium production is the constant measurable decay of uranium.

If earth is billions of years old the atmosphere should be saturated with helium! There should be up to a million times more helium than we have now!

Using helium as an indicator of age is complicated by a number of unknown factors related to losses of earth's atmosphere. At present our atmosphere is approximately half of what some believe to have existed prior to the flood.

According to some experts, the helium "clock" insists that the earth cannot be more than 10,000 to 15,000 years old.[30]

Do Changes in Our Planetary Neighborhood Tell Us More?

The Moon Is Receding

Scientists observe that tidal friction and other things are making the earth's rotation speed slow down a very tiny amount each year. Though it is not significant enough to make a serious difference on the earth, even over a few billion years, it does result in another interesting effect. The moon's distance from the earth is constantly increasing!

Accurate recent lunar measurements show the moon is drifting two inches farther from the earth each year. The rate would have been higher in the past. Though that may not sound like much, if you believe the evolutionary view that the moon has been circling the earth for five billion years, you have a problem. Working backwards, the moon could not have been orbiting the earth for more than 1.4 billion years. In five billion years the moon would have receded 380,000 miles and would have significantly different effects on the earth, like much smaller tides.[31]

The Sun Is Shrinking

A news wire report of March 23, 1980 made the surprising statement:

"The sun's diameter appears to have been decreasing by about one tenth percent per century!"

Scientists have been watching for over a hundred years. Additional studies have shown that the rate is variable, but on the average it is more like four hundredths of a percent per century. **Every hour the sun is shrinking about two feet.**

Of course two feet an hour isn't much when you consider the sun is nearly a million miles in diameter (840,000 miles). But what are the implications?

If the sun is shrinking 0.04 percent per century, then it totals 0.4 percent per millennium.

Years ago, many people believed the sun was made of burning coal. The idea was rejected because the sun could not have lasted more than 5,000 years if coal was the source of its heat. In the 19th century, a German scientist named Helmholtz suggested that the sun's energy was produced because it was shrinking. This was a perfectly sensible idea. The sun could keep burning for more than 20 million years, but since many scientists had begun to believe the earth was several **billion** years old, they wouldn't accept this explanation. Instead, they suggested the sun was a huge nuclear reactor, producing energy by turning hydrogen into helium through a process called nuclear fusion. If this is really happening inside the sun, there should be trillions of tiny (sub-atomic) particles called neutrinos racing away from the sun and passing through the earth. Special tests found hardly any neutrinos. Believers in the fusion theory were puzzled so they invented the idea that neutrinos changed their nature and can't be detected. Recent close-ups of sunspots show that the sun's interior is cooler than the surface where there is terrific electrical activity, some in the form of huge plasma filaments. Senior science writer Robert Britt wrote, "So, without any direct evidence that the thermonuclear powered model of the Sun is correct, and with strong evidence against it, we should...doubt it."[32.1]

Now if you believe the sun's age is only 6,000 years there's no real problem. In that time the sun will have shrunk only about 2.4 percent. Life on earth would go on quite fine. But if you believe the earth and the sun have been around for nearly five billion years, you've got a problem!

If the sun existed only 250,000 years ago it would have been double its present diameter. At that size with the earth at its present distance from the sun, it would be too hot on Earth for life to exist.

Thirty million years ago the sun's surface would have been touching the earth!

As far as researchers can measure, there is no reason not to think this rate of shrinkage, or something close to it, has been consistent since the origin of the sun. But astronomers also admit that stars much larger than the present size of our sun, burn hotter and shrink faster than the sun. From the pure simple evidence it is clear that life would have been totally impossible on earth even a million years ago. Or perhaps the sun and earth just aren't all that old![32]

"They shall perish, but thou shalt endure: yea, all of them shall wax old like a garment; as a vesture shalt thou change them, and they shall be changed."
Psalm 102:26

Three thousand years ago a Psalm was written that says: *"I will lift up my eyes to the mountains; from whence shall my help come? My help comes from the Lord Who made heaven and earth!"*
Psalm 121:1-2

How old is the universe?

Astronomers using big-bang cosmology have calculated the age of the universe to be 15 to 20 billion years. They often refer to a phenomenon observed in stars called "red shift." They assume that red shift is a result of velocity (a reasonable initial assumption). By correlating red shift with relative brightness of stars and galaxies, they have built a model of a supposedly expanding universe, with distant objects billions of light years away. It is unfortunate that the public is seldom made aware that not all the experts agree.

Recent studies indicate that red shift is an intrinsic property of stars and galaxies and that it has little bearing on their distance OR velocity. A general conclusion from these studies is that the universe is neither expanding nor contracting.[33]

Age? The amount of spiral in spiral galaxies is used to estimate ages of millions of years, but the original conditions of the galaxies are not known. These numbers are pure speculation. Other models suggest that the spiral appearance of existing galaxies could no longer exist if they had been spinning for the billions of years popularized by many.

What Can We Learn From the Universe about Its Origin?

Ideas about the origin of the universe are always speculation. Since no human was there to observe and report about it, we must be honest to admit that statements about supposed processes of the past are unproven and unprovable. All we can do is ask: "What do present processes suggest about the dim past of the cosmos?"

No astronomer has ever seen a star born, much less evolve into a more complex structure!

Astronomers HAVE SEEN the violent destruction of a few stars. They see stars associated with dust clouds. Theories about origins are sheer speculation.

Degeneration and disintegration is certainly not evolution.

In astronomy, a great deal of importance is placed on SEEING things. It is extremely faulty "science" to announce a process that cannot be observed. Yet astronomers talk and write often about stars being born as if they were sure of it, and positive that it happened through millions of years of time.

Note: *National Geographic*, Jan. 1985, page 8 states:

> "...new stars are still being born - as they always will."

THINK! Since no one has ever seen such a birth, how much honesty is represented in a statement like that?

SPHERES OF INFLUENCE

All stars, planets, comets and asteroids have their own local gravitational fields. It is commonly assumed that particles like gas, dust and meteors that come within the sphere of gravitational influence of larger bodies will be pulled into those bodies. But this is not what is observed.

An example is found in the space program where the astronauts on the way to the moon dumped waste material outside their spacecraft. Where did it go? It didn't go off into space, and it didn't fall back onto the spacecraft. It followed the spacecraft all the way to the moon slowly orbiting the spacecraft as captive moons.

Planets cannot "sweep up" material objects moving in orbits similar to their own. An object outside a body's sphere of influence cannot reach the edge of the sphere if it doesn't reach a relative velocity that matches escape velocity at that distance from the body. In other words it has to be moving very fast toward the body. So such an object cannot be reasonably captured (unless forces other than gravity are involved).[37]

> O LORD, Our LORD, how majestic is Thy Name in all the earth, Who hast displayed [set] Thy splendor above the heavens! When I consider [see] Thy heavens, the work of Thy fingers, the moon and the stars, which Thou hast ordained [appointed, fixed]; what is man that Thou dost remember him?
>
> *Psalm 8:1,3*

Conditions in our solar system that don't fit evolutionary expectations.[38]

1. Suns and planets do not condense from cold clouds of gas and dust.

2. The sun has a very small angular momentum compared to the planets (1/200th).

3. The system's major angular momentum is in the planets.

4. There are eccentric and even tilted planetary orbits.

5. Uranus and Venus rotate in the opposite direction of the rest.

6. Some of the planets' satellites are also in retrograde motion.

7. The angular momentum is evenly distributed among the planetary satellites.

8. Our moon has a lower density than the earth.

9. The heaviest elements are predominantly in the smaller planets and closer to the sun.

10. There is evidence to indicate that some earth rocks were formed as cold, hard material.

The heavens are the heavens of the LORD, but the earth He has given to the sons of men.

Psalm 115:16

Is there evidence to falsify the big-bang theory?

First, draw a picture of NOTHING, then show it EXPLODING!!

Consider some observations that don't line up with a super-explosion cause for the universe 10 - 15 billion years ago:

1. **Supernova remnants** (debris from an exploded star) should have tallied in the thousands by now, if the universe is old. Only 205 have been detected. This number is 65 less than expected by astronomers, even assuming the universe is only 7,000 years old.[34]

2. **Deep space galaxies appear "mature"** (where they should be young) and much the same as closer galaxies, indicating galaxies have not evolved, but were all created much as we see them today.[35]

3. **Red dwarfs are assumed to be faint old stars that should number in the thousands** if the universe is billions of years old. However, astronomers have reluctantly admitted that the limited number found fits a biblically young universe of 10,000 years or less.[36]

Are Radiometric Dating Methods Reliable?

"Test and prove all things."
1 Thessalonians. 5:21 (Amplified)

Let's ask some questions about those dating methods that supposedly give us absolute ages for the earth in the millions and billions of years.

When you pick up a volcanic or igneous rock, is there really some honest scientific way to figure when that rock cooled down? Museums tell us they have highly technical ways to analyze the elements and determine how long ago these types of rocks were formed. These are the radiometric dating systems that have only been used for the last few decades. Many people have been indoctrinated to think these are authoritative methods to establish absolute ages for rocks, but are they truly realistic?

You've probably heard about potassium-argon dating or uranium-lead dating. Each is based on several assumptions about the decay process of an unstable element. For example, radioactive uranium (an unstable element) breaks down very gradually to form the very stable element lead. The decay process is very slow, supposedly taking billions of years to transmute even half the original amount of uranium into lead.

If you find a fossil buried beneath a layer of rock that can be radiometrically analyzed, you should get a fair idea of the fossil's age. But remember: there are a bunch of assumptions in all this.

Assumptions In Radiometric Dating

1. No daughter element was present at the start of the process.

2. The decay rate has always been the same.

3. There were no changes in mineral content from percolation or leaching.

With the exception of the carbon-14 test, it's important to remember that these few processes can only be observed in igneous rocks. Thus volcanic rocks (usually) are tested to yield the supposed age of a fossil or artifact found buried beneath them.

Do radiometric tests really check out?

Apollo 11 rock samples produced a variety of results when tested with the uranium-lead method: 4.6 billion years, 5.4 billion years, 4.8 billion years, and 8.2 billion years. Another report gave a potassium-argon age of 2.3 billion years.[39]

What is the acid test to see if these "absolute" dating clocks are really telling the right time?

The best way to test a clock's accuracy is to compare it to a "standard." In other words check with a known reliable source.

Did you know? There is another way to know the age of a volcanic rock!

Case History #1

Volcanic lava rocks from Hawaii were subjected to potassium-argon testing. Result? Between 160 million and 3 billion years! That's how long ago this lava supposedly cooled down from a molten state and the potassium "clock" was set. [Now stop and think about that. How much flexibility does science need when it comes to estimating numbers like this?]

Upon further checking it was discovered that the particular lava flow from which these rocks were taken, actually erupted in the year 1801![40]

THINK! Do you suppose there just might be some flaws in the radiometric dating methods?

Case History #2

Tests were made on volcanic rocks from Russia with results ranging from 50 million to 14.6 billion years old. Historical research determined that these very rocks had actually erupted only a few thousand years ago.[40a]

Is volcanic rock age testable?

The following familiar discoveries were all "dated" by the radiometric tests on volcanic (ash) material overlying the actual bone samples.

1. "Skull 1470" was publicly proclaimed to be 2.8 million years old. They got that from the potassium-argon dating of the volcanic ash in which the skull was buried as reported in *National Geographic* in June 1973.

2. "Lucy" is a discovery of skeletal fragments claimed to be three million years old. Again, they got that from the radiometric date on the volcanic ash in which it was buried as reported in *National Geographic* in December 1976.

3. Fossilized human footprints were also found in Africa, in a layer of volcanic ash claimed to be 3.6 million years old according to *National Geographic* in April of 1979.

CRITICAL QUESTION

When there are discrepancies over a rock's age when the actual age is known, do you suppose there might be similar discrepancies regarding rocks of totally unknown age?

Can we assume these ages are correct when the only tests performed were made on the volcanic ash overlying the objects?

Other famous finds have produced some helpful added data.[41]

When Australopithicus was found in Ethiopia, it was publicized to be one to two million years old, because of the potassium-argon dating of the overlying volcanic ash. Since there were many expendable animal bones also found alongside, these were carbon dated directly. They yielded a carbon age of just 15,500 years old!

In Kenya, Africa, the bones of Zinjanthropus were hailed as two million years old because of the potassium-argon date on the overlaying volcanic ash. Again, lots of other animal bones were found in the same deposit that carbon tested at just 10,000 years![42]

These facts are freely available in technical science journals, but why don't we hear about these contradictions in the popular magazines?

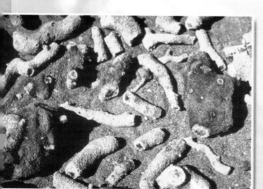

SURPRISING DISCOVERY

In 1973, a broken power line caused the petrifaction of many tree roots in the sandy soil near Grand Prairie, Alberta in Canada. What happened? The heat was so intense that the wood just vaporized, leaving hollow molds in the ground. As the electric current baked the surrounding soil it got so hot that some of the silicon dioxide (sand) literally boiled. Some of the intensely hot silicon vapor found its way into the root molds where it re-crystallized filling the place of the roots. That's what you call instant fossils!

Some of the roots were taken to a Canadian university where the earth scientists were asked: "What kind of a potassium-argon date would you get on these roots?" They said, "The test would be meaningless. It would indicate an age of millions of years because heat was involved in the petrifaction process."[43] If heat renders the radiometric tests meaningless, how can tests on volcanic ash be helpful to establish when it was formed?

Is Carbon-14 Testing Reliable?

What is it?

In 1947, Willard Libby found how to measure molecular amounts of unstable carbon-14 in old objects. The test supposedly reveals how long ago something died.

How does it work?

As cosmic radiation enters Earth's upper atmosphere, a series of reactions causes some nitrogen-14 atoms to change to carbon-14 (C-14). The carbon isotope is unstable. About half these unstable atoms decay back to stable nitrogen-14 in about 5,730 years. Carbon (both isotopes) unites with oxygen molecules in the atmosphere to form carbon dioxide. By photosynthesis, plants everywhere receive carbon dioxide, some of which contains C-14. And as animals consume the plants, minuscule amounts of new C-14 are incorporated in their bodies. So all living things are expected to contain C-14. Then when a plant or animal dies it stops receiving new C-14, and all the C-14 present in the specimen at that time continues breaking down. So all living things are expected to contain C-14.

What are its limits?

The "half-life" of C-14 is only 5,730 years. In five half-lives (29,000 years) very little C-14 isotope remains. An accurate age is impossible to know. [A "half-life" is the time required for half the unstable element to break down into its "daughter element"]. If an object is presumed to be older than 50,000 years, a C-14 lab won't bother with it.

Worse yet, scientists, including Dr. Libby, acknowledge that C-14 production exceeds the decay rate by about 25%. Yet the system assumes that decay and the formation of C-14 are in equilibrium, which would evidently be reached in 30,000 years of rather consistent environmental conditions with no major catastrophic changes.[44]

The assumptions of Carbon-14 dating:

1. The formation in the atmosphere has been constant for at least 70,000 years.

2. The formation is the same all over the world.

3. The content ratio is the same in all kinds of specimens worldwide.

4. Ancient specimens are not contaminated with solutions containing modern amounts of C-14.

5. There is no loss of C-14 except by radioactive decay.

Is there room for error?

The C-14 system depends on a steady, unchanging rate of radiation in the atmosphere for the last 30,000 years. The unproven assumptions make the system undependable.

Let's illustrate this. Imagine appearing in a closed room with only a candle burning and you are asked to figure out how long the candle has been burning. As you think about it, you realize the challenge is impossible. You might figure how long the dripping wax took to accumulate. Or you could get technical and try to measure the relative amounts of oxygen and carbon dioxide gases in the room compared to outside. But how would you know if someone had opened and re-shut the window? Could any past condition have caused it to burn faster? Was it ever put out and relit?

What things can alter the generation rate of Carbon-14?

1. Decay of the earth's magnetic field. As the field weakens, more cosmic rays enter the upper atmosphere resulting in an increase in the generation rate of C-14.

2. Air pollution from volcanic activity and industrial burning can shield solar input and alter gas ratios in the air.

3. Changes in solar activity. Solar flares and sunspots can cause a temporary increase in the generation rate of C-14.

4. Changes in cosmic radiation levels reaching the upper atmosphere from extraordinary events in our galaxy like a supernova (explosion of a star).

5. Meteors or asteroids falling to earth.

How can meteors affect C-14 concentrations? Read about the "Riddle of the Great Siberian Explosion" in *Reader's Digest,* August 1977. It happened on June 30, 1908. The carbon-14 measurements of tree rings around the world were evidently altered as a result of this single blast.

If the generation rate of C-14 was less in earlier times, then the C-14 level in the atmosphere would have been less. Specimens from long ago would carbon date older than their real age because they contained less C-14 from the start.

How can we test ANY sample and estimate the effects of past environmental changes? The C-14 system demands we assume there have been NO globally catastrophic events in the past 50,000 years. If conditions on earth were very different in the past, and especially before the Flood, then C-14 is nearly worthless, particularly for ages assumed to be older than 5,000 years.[45]

What makes C-14 dependent on geographical location?

Some organisms don't get all their carbon from the air. Carbon from water and soil can have low amounts of C-14, making them look old by C-14 standards. For example:[46]

1. Seal Skins (fresh) dated at 1,300 years old.

2. Living mollusks dated at 2,300 years old.

3. Living plants growing by springs dated at 17,300 years old.

The atmosphere in certain locations may have high concentrations of old carbon (C-12) due to burning of hydrocarbon fuels or from volcanic activity. For example: A living tree growing next to an airport dated at 10,000 years old. Carbon dioxide from aircraft engines diluted the C-14 concentration in the air.[47]

What causes contamination?

Older samples are more subject to contamination than recent ones because they have smaller amounts of C-14. Wood and charcoal samples in soils where water is percolating can gain or lose C-14, depending on the C-14 level in the water. Museum samples kept in wooden cases or wood fiber containers can gain C-14 from their containers, thus altering their original radiometric age.

The dating system is based on uniformitarian philosophy and long ages, believing environmental conditions have remained constant for many thousands of years. How does that concept fit into biblical history that tells of a young earth with catastrophic changes to the earth?

Can C-14 be used for dating anything with reasonable accuracy?

Items less than 3,000 years old can be dated for comparison with reasonable accuracy when one understands the limitations. Comparing like kinds of samples will give the best determination of age differences. But many thousands of "old" things have been C-14 tested with results that are shocking to those accepting the concept of long ages.

Do Radiometric Dates Verify The "Approved" Geologic Chart?

According to the popular traditions of our day this chart is intended to portray the progressive "ages" of Earth's history.

It is admitted that the carbon-14 (C-14) testing system, when calibrated to account for the lack of equilibrium between atmospheric generation and terrestrial decay, can be useful in determining the ages of items in the lower range of 3,000 years or so. However, many thousands of things have been C-14 tested, some with very shocking results. The following information comes from several issues of a scientific journal called *Radiocarbon*. Keep in mind that things older than 50,000 years could not have enough C-14 to measure even if they really were that old.

Look at this surprising admission by an evolutionary scientist, "The troubles of the radiocarbon dating method are undeniably deep and serious.... Continuing use of the method depends on a 'fix-it-as-we-go' approach,

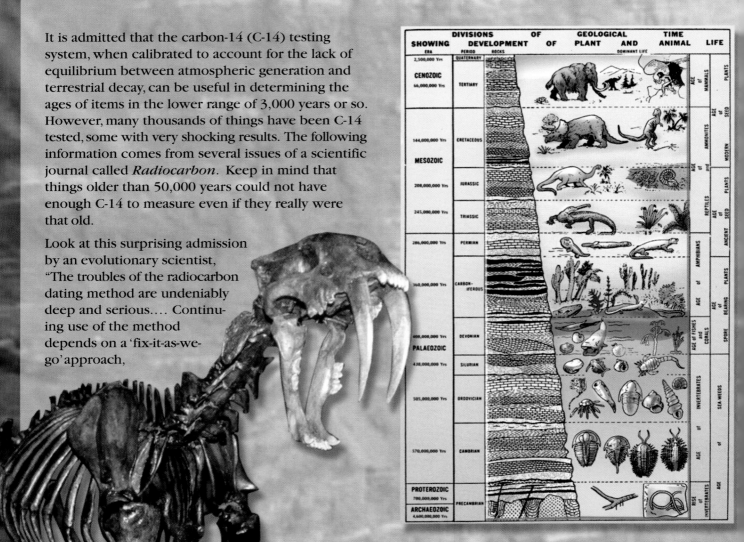

OBJECT TESTED	SUPPOSED AGE (BY THE GEOLOGIC CHART)	C-14 TEST RESULT
Saber-toothed Tiger	100,000 – 1,000,000	28,000
Natural Gas	50,000,000	34,000
Coal	100,000,000	1,680

allowing for contamination here, fractionation there, and calibration wherever possible. No matter how 'useful' it is, though, the radiocarbon method is still not capable of yielding accurate and reliable results. There are gross discrepancies, the chronology is uneven and relative, and the accepted dates are actually selected dates."[48]

Are radiometric tests used to determine the validity of assigned ages on the standard geologic chart?

ANSWER: *NO*

CHALLENGE: If the geologic chart offers any factual support to assumed evolutionary ages, and if radiometric dating methods have any absolute scientific accuracy, why won't evolutionists undertake a complete analysis of the entire sequence?

Keep in mind that the "geologic eras" were named over a century ago to identify rocks formed before, during and after the Flood. Since then, the names have been largely used by evolutionists to associate evolutionary ages to rocks and fossils. They mean only what believers in the theory want them to mean. The names of the "periods" reflect locations where key fossil deposits were found. The names were assigned long before radiometric methods were even known.

Consider the following:

"Paleo-zoic" means ancient life.

"Meso-zoic" means middle life.

"Ceno-zoic" means recent life.

"Cambrian" refers to the ancient name for Wales in Great Britain.

"It is obvious that radiometric techniques may not be the absolute dating methods that they are claimed to be. Age estimates on a given geological stratum by different radiometric methods are often quite different (sometimes by hundreds of millions of years). There is no absolutely reliable long-term radiological 'clock.' The uncertainties inherent in radiometric dating are disturbing to geologists and evolutionists."[49]

One geologist was honest enough to admit the shortcomings of the whole system by writing, "No coherent picture of the history of the earth could be built on the basis of radioactive datings."[50]

Even carbon tests on freshly harvested clams have yielded results suggesting that over 2,000 years have passed since they died.

Evolution demands millions of years. But if millions of years are mythical and evolution is untrue, then the whole foundation of historical geology vanishes in the tide that destroys every house built on error.

So, is the earth old or young?

Can the Rock Layers of Earth Be Interpreted Another Way?

In our day, when most people admire the massive layers of mountain formations, they tend to automatically think in terms of millions of years. We've been programmed that way. The charts in the parks we visit were designed with evolution in mind. Our indoctrination in materialism insists that gradual processes of erosion and mountain building happened so slowly that only hundreds of millions of years could possibly account for it all.

But what really formed the rock layers?

Around the world you commonly see three main features exposed on the geographical surface of our planet.

The first feature we find is sedimentary rock. Sedimentary layers of sandstone, limestone, and shale are found all over the world. They were clearly deposited by moving water. These layers continue in every direction for many miles. Some of them are only inches thick, while others are hundreds of feet thick. For the most part, our entire planet is overlaid by up to several miles deep of layer after layer of sedimentary rock.

Secondly, volcanic rock layers also extend over much of the earth. Often they are interspersed among the sedimentary layers; and often with evidences of vast volcanic eruptions under water.

Lastly, fossils are found embedded abundantly in the rock layers. Plants as well as all kinds of animals are jammed into mass graveyards all over the earth. Have you ever paid particular attention to the condition of their burial? It can tell you a lot. Geologic fossil formations virtually always show evidence of catastrophic burial. Jumbled piles of bones, plants and mixtures of various types of creatures tell us a gruesome story.

When you find a fossilized fish with its mouth open wide, "frozen" in the act of swallowing another fish, what does that tell you about its burial? Could it have happened slowly? No way! Huge regional graveyards of sometimes millions of creatures were instantly entombed in billions of tons of mud.

At Petrified Forest in Arizona, untold thousands of petrified giant sequoia trees are eroding out of layers made of flooded volcanic ash. It's a massive burial that could not have happened slowly.

In Alberta Canada you can see the valley of the Red Deer River cutting through thousands of square miles of catastrophically flooded volcanic ash. It's loaded with dinosaur bones! Literally millions of fossils here testify to a horrifying flood disaster. Over 200 whole dinosaur skeletons have been excavated here and sent to museums around the world. Climbing around the strange formations is a lot of fun, and it's amazing how many fossils you can find just lying around.

Petrified skeletons of ordinary fish like herring and cod are found by the millions in huge formations of soft shale extending over much of the state of Wyoming. These fish literally choked to death in a flood of mud covering a vast region now sitting over 3,000 feet above sea level.

How does the Bible explain all this?

According to Genesis, the Flood was a worldwide cataclysm lasting over a year. The Flood is the most historically noted event of ancient time. Over 300 non-biblical, historical accounts from tribes and nations around the world claim the cataclysm was real! But why don't we hear about it in World History 101?

With a massive upheaval like the Great Flood of Noah's time described in the Bible, we'd expect to find evidence of:

1. Vast sedimentation on a global scale.

2. Widespread volcanism.

3. Fossilization globally.

Think what a single volcanic eruption, tidal wave or local flash flood can do!

1. In seconds whole towns are swept away and buried.

2. House-sized boulders are moved miles downstream.

3. Sediment dozens of feet thick can accumulate in minutes.

Doesn't it become clear what enormous potential destruction there is in a worldwide flood?

The rock layers, mountain upthrusts, and many erosional features of Earth are easily accounted for by the Flood. The problem with many people's thinking is that they make the Flood too limited when, in fact, it was the most awesome destruction our planet has ever endured.

The imaginary geologic chart with its "millions-of-years-long" epochs is totally meaningless if a worldwide flood really did happen. Could it be that maybe it didn't take millions of years to make huge geologic formations?

After seeing the many evidences we've considered in this section, one fact becomes exceedingly apparent...

THE EARTH IS YOUNG!

Is There a Gap of Time in Genesis?

What Is This "Gap" All About?

Many conservative Christian scholars of the 20th century adopted an unorthodox interpretation of the Genesis narrative of creation. This is called the "gap theory" by many. According to this theory, a large period of unrecorded time, possibly millions of years, followed what is assumed to be the first completed creation described in verse one of the first chapter of Genesis. This unwarranted interpretation of scripture is only taken in order to accommodate the evolutionary millions of years, which many assume to be legitimate, even though the scientific evidence to support them is insufficient.

The description, "without form and void" in Genesis 1:2 is taken to mean that the entire creation became ruined and destroyed prior to the virtual "re-creation" described through the rest of chapter 1. The supposed destruction of the world at this time is said to have been the result of the fall of Satan. The theory maintains that the entire earth was filled with plants and animals, and that even a race of "pre-Adamic" men ruled the earth. Then, with Satan's rebellion, darkness invaded the perfect earth and a global flood destroyed everything.

The vast ages of the evolutionary geologic chart continued during this time gap. The fossilized plants and animals of our earth's rock layers are supposed then to be the remains of that originally perfect earth, destroyed prior to the six literal days of *re-creation* recorded in Genesis.

Supporters of the gap theory say that Genesis 1:28 hints that this whole story may have happened. The King James Version reads that God told the first human couple to "be fruitful, multiply and replenish the earth..." The thought here is that a destroyed civilization had to be replaced and the earth REFILLED with inhabitants. However, the word "replenish" was used in the King James because the word "plenish" was not a word in use to signify the meaning "to fill." Today, many of the modern translations just use the word "fill" in Genesis 1 verse 28.

Where Did This Idea Originate?

Actually it goes back to the year 1814. The budding science of geology in Great Britain was fostering new ideas about the age of Earth's geologic features. Scottish theologian, Dr. Thomas Chalmers, proposed the "ruin-reconstruction"

idea to seemingly make room for long ages of time in the Genesis record. Through the course of the 1800s, evolutionary thoughts about the extreme age of the earth and universe were increasingly promoted. The public became indoctrinated with the idea that the earth was millions of years old. The straightforward historical and orthodox view of the biblical account was under fierce attack in academic circles.

The gap theory was further popularized with the publication of George Pember's book, *Earth's Earliest Ages*, in 1876. But the idea really gained wide acceptance after it appeared in the footnotes of the Scofield Reference Bible in 1917.

For a more thorough study of the gap theory you should get a copy of John Whitcomb's book, *The Early Earth* as well as Ian Taylor's *In The Minds Of Men*.

Now, we realize that there are many good, sincere Christian scholars – men and women called to serve God – who have accepted and taught the gap theory as true. Though several Bible passages are used to support the idea, with increasing public attention on the origins debate, the gap theory is no longer taken seriously by most conservative Christian scholars. Let's be honest enough to realize that it is only a theory. Let's also recognize there are some genuine difficulties with it.

Analysis of Gap Theory Problems

Let's analyze the problems raised by the gap theory.

1. A major problem arises before you even get out of Genesis 1. In verse 31, God saw ALL that He made and "behold, it was VERY GOOD."

- What does "very good" mean? Had the earth been destroyed leaving billions of fossilized animals as evidence of the destructive judgment? Could such a world legitimately be called "very good?"

Was Adam created to walk over a virtual graveyard filled with all manner of life forms which God had created and subsequently destroyed? This brings up another problem.

2. When did death begin?

- According to the Bible, did death on this earth begin with some Satanically ruled race?

- Romans 5:12 plainly says: "...by one MAN sin entered into the WORLD, and death by sin..."

- First Corinthians 15:21 clearly relates to Adam, the first man, and the only "first" man the Bible ever acknowledges. It says, "by a man came death."

- Even those who believe the gap theory admit that the first few verses of Genesis are clearly lacking a context of judgment and death.

- Keep in mind that the supposed flood of destruction following eons of time did not occur until the fall of Satan. The Bible shows that the "groaning and travailing in pain" of the animal kingdom is a result of the curse brought on by Adam. How then could animals have died for millions of generations even before Satan's fall?

That brings up another problem.

3. When DID Satan fall?

- We know that when Satan was created he was good and without rebellion. When was Satan created?

- In Genesis 2:1 it says: "Thus the heavens and the earth were completed, and all their hosts." Exodus 20:11 elaborates further: "...in six days the Lord made the heavens and the earth, the sea and all that is in them." Other Bible contexts refer to the all-inclusiveness of the creation event. Nehemiah 9:6 is an example. Here it says: *"Thou hast made the heavens, the heaven of heavens with all their host, the earth and all that is on it, the seas and all that is in them."*

- Some may say that such an idea doesn't give a long enough existence for Satan before Adam's creation. We simply have to ask the question: Is such a long time required biblically?

- After all, how long after his creation did Adam disobey God by eating of the tree of knowledge of good and evil? Probably not a very long time at all. We know that all of Adam's days totaled 930 years. His first commission was to be fruitful and multiply. Eve's first child would no doubt have been conceived within just weeks or even a few days after her creation. [The Bible calls Eve the *"mother of all"* humans, and we also realize that *"all (humans) have sinned..."*][49] If Adam fell in sin in such a short time, do you think it's possible that Satan also could have fallen in a relatively brief time after his creation?

- When God saw all He had made, was it ALL very good? If Satan had already fallen by the sixth day of creation, was God just ignoring him in this statement? Did God really mean *"everything I made is good except for Satan and his rebellious angels?"*

4. Whom did God create to be the dominator of this earth anyway? Does the Bible ever indicate that anyone else but Adam was specifically made to rule this earth?

- Genesis 1:26 says: *"...Let us make man in our image...and let them rule...over all the earth."*

- Were there ever animals made by God (like dinosaurs) over which Adam did not have dominion? Some suggest that Satan created some of the grotesque giants of the prehistoric past. But does Satan have the power to create life? Or is he the one ultimately responsible for the perversion of life? Does the Bible say anything about a demonically linked corruption of Earth's animal life?

5. And what about the Great Flood of Noah's time? Does the gap theory maximize the impact of a pre-creation week flood while minimizing the biblically historical flood of Genesis?

How Devastating Was the Genesis Flood?

The supposed flood of the gap theory left the earth "without form and void." Thus, many have been falsely led to believe that the majority of our planet's geologic history took place before the creation of Adam and the six-day creation account in Genesis.

Here's what that would mean. All the major fossil-bearing rock formations, the miles of thick strata like we see exposed in the Grand Canyon, and many of the mountains we see today, would supposedly have been produced by an awesome catastrophe summed up in less than a dozen words of Genesis 1 verse 2.

Think ! If we take that approach, we really create a problem in understanding the destructive influence of the Genesis Flood of Noah's time. We have several choices:

1. Was the Flood of Genesis 6, 7 and 8 merely a **local flood**, relatively insignificant in terms of geologic formations? Many traditional (global) flood legends from widespread cultures (from Hawaii to Persia) strongly contradict this choice.

2. Was it a **worldwide flood affecting only surface features**? H.H. Howorth's *The Mammoth and the Flood* documents in detail portions of the earth that have surface fossil deposits produced by a short term event of rapidly moving water. This would seem to be logical evidence for a post-(Noah)-flood catastrophe that includes significant volcanic and earthquake evidences.

3. Must it have been a **worldwide flood** responsible for **all major sedimentary deposits and fossils?** Although a popular concept, it does not recognize evidence from a post-flood event.

4. Could it have been a **worldwide flood** responsible for most of the major sedimentary deposits, but was **followed by**

post-flood events that were responsible for many of the surface features?

If the Flood of Noah was not a local flood or a short-term event producing only surface deposits, it must have been an earth-shattering, yearlong catastrophe that totally rearranged the structure of Earth's crust. Then evidence of a previous destruction (at the end of some supposed "gap" of time before the completion of the "very good" creation) would most likely have been obliterated.

Further study of the Genesis Flood is really essential to properly understand foundational realities of the Bible, world history, and natural sciences. Contact creation ministries like the Institute for Creation Research and the Creation Resource Foundation for further recommended resources on the subject.

All this discussion really brings us back to where we started: When did it all begin? If the earth really is only a few thousand years old and the present land-forms were produced by the awesome destruction of Noah's Flood, then we don't really need a gap, do we?

The opening verses of Genesis describe not a destroyed earth, but merely an incomplete earth. The evening and the morning marked the first day in God's masterful building project. Let's always remember that God's Word is not some deep dark mystery. But as we already learned in Proverbs 8:8 and 9, God's words are straightforward. They give understanding even to the simple as it says in Psalm 119:130.

Indeed, the Bible is the key to unlock the mysteries of Creation. Be careful not to program your computer with unproven theories as if they were absolutely true.

Remember that God is not afraid of the facts. Every idea bears inspection. That's why God dares to say in First Thessalonians 5:21, "Prove all things, hold fast to that which is good."

The same thing applies to our ideas about salvation and our eternal existence. Be careful not to take chances about the things of God and your spirit. What you believe is the foundation upon which you build your actions.

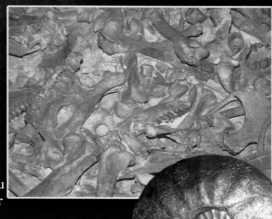

The words of Jesus have withstood the test of time. How and why is it that faith in Him alone can guarantee a restoration of relationship between you and God?

His Word tells us in Colossians chapter 1 that it's because of Jesus that we can be transferred out of the kingdom of darkness and into the kingdom of light. How can Jesus have so much authority? Verse 16 tells us it's because all things were created through Him. In fact the next verse says that in Jesus everything holds together. The secrets of the universe are not ever going to be found in science alone. No mere man or guru has all the answers. Think of what the Creator has demonstrated to us through Jesus Christ. His Spirit is sovereignly declaring that reality to us now.

> "... all things were created by him, and for him. And he is before all things, and by him all things consist."
>
> *Colossians 1: 16-17*

Unlocking the Mysteries of Evolution
It's time to set the record straight

"Any story sounds true until someone tells the other side and sets the record straight."

Proverbs 18:17 TLB

What has set the stage for the present condition of popular modern thought?

After several generations of increasing materialism in our culture, the vast majority of people seem to be greatly ignorant of the faith of their forefathers. Many college-age, anti-establishment libertines of the hippie generation in the late 20th century have become the university professors of the early 21st century. School students who have been conditioned by irreverent rationalism and skepticism have been increasingly indoctrinated with humanistic ideals rather than the One true God of our culture's Christian ancestors. The general public is typically led to believe the so-called scientific declarations of intellectuals who have thrown off the restraints of a God-honoring mindset.

By the inspiration of the Holy Spirit, the apostle Peter made a remarkably accurate prediction of the conditioned human thinking we find prevalent today.

> "Knowing this first, that there shall come in the last days scoffers, walking after their own lusts and saying, 'Where is the promise of his coming?' for since the fathers fell asleep, all things continue as they were from the beginning of the creation. For this they willingly are ignorant of, that by the word of God the heavens were of old, and the earth standing out of the water and in the water: Whereby the world that then was, being overflowed with water, perished. But the heavens and the earth, which are now, by the same word are kept in store, reserved unto fire against the day of judgment and perdition of ungodly men." *2 Peter 3:3-7*

Notice the driving influence of these scoffers. It is willful ignorance! And the truths that they are specifically ignoring are: 1) the creation, 2) the global Flood, and 3) the reality of the judgment to come by fire. Isn't it interesting to see that the underlying creed that generates that willful ignorance is an adherence to what seems to be a uniformitarian view that "all things continue as they were from the beginning"? Let's see why the evolutionary presupposition has gained such a strong hold on modern thinking.

ORIGINS: By Dr. Wernher von Braun

(the leading founder of the modern space exploration program called NASA)

One cannot be exposed to the law and order of the universe without concluding that there must be design and purpose behind it all.... The better we understand the intricacies of the universe and all it harbors, the more reason we have found to marvel at the inherent design upon which it is based.... **To be forced to believe only one conclusion – that everything in the universe happened by chance – would violate the very objectivity of science itself**.... What random process could produce the brains of a man or the system of the human eye?... They (evolutionists) challenge science to prove the existence of God. But must we really light a candle to see the sun?... They say they cannot visualize a Designer. Well, can a physicist visualize an electron?... What strange rationale makes some physicists accept the inconceivable electron as real while refusing to accept the reality of a Designer on the ground that they cannot conceive Him?... It is in scientific honesty that I endorse the presentation of alternative theories for the origin of the universe, life and man in the science classroom. **It would be an error to overlook the possibility that the universe was planned rather than happening by chance.**

Dr. von Braun was a leading scientist and outspoken Christian influence in the NASA space program of the United States until his death in the late 1970s. These remarks from a personal letter, were read to the California State Board of Education by Dr. John Ford on September 14, 1972 [1]

Who Has Insight... Let Him Hear...

Can an atheist understand the battle better than most Christians? Note what one wrote:

> Christianity has fought, still fights, and will fight science to the desperate end over evolution, because evolution destroys utterly and finally the very reason Jesus' earthly life was supposedly made necessary. Destroy Adam and Eve and the original sin, and in the rubble you will find the sorry remains of the son of god [sic]. Take away the meaning of his death. If Jesus was not the redeemer who died for our sins, and this is what evolution means, then Christianity is nothing.

Christianity, if it is to survive, must have Adam and the original sin and the fall from grace, or it cannot have Jesus the redeemer who restores to those who believe what Adam's disobedience took away.

What all this means is that Christianity cannot lose the Genesis account of creation like it could lose the doctrine of geocentricism and get along. The battle must be waged, for Christianity is fighting for its very life.[2]

— published by an American atheist periodical

The Roots of Evolutionary Faith

"Know this first of all, that in the last days mockers will come with their mocking, following after their own lusts, and saying, 'where is the promise of His coming? For ever since the fathers fell asleep, all continues just as it was from the beginning of creation.'"
2 Peter 3:4

Until the mid-1800s, brilliant scientists had no trouble accepting the biblical account of creation. Remember Isaac Newton, one of history's most ingenious scientists? Other familiar greats include Kepler, Faraday, Kelvin, Mendell, Lister, and Pasteur. They all lived and died with faith in God, but a new generation arose during the 19th century in the midst of much skepticism and moral decline in society. What happened?

UNIFORMITARIANISM

Charles Lyell, a young attorney, became one of the pillars of modern geology. He published his *Principles of Geology* in three volumes from 1830-1833. It has literally shaken the intellectual world. The study of Earth's geophysical structures was an infant science then, but now his concepts are so widely accepted that most who believe them don't even know who originated them.

Lyell's basic premise came to be adopted by all scientific fields, even though his field of study was

earth science. He proposed that all our earth's features can be explained in terms of present observable processes, which supposedly have always gone on in the past at the same rates as today. Thus:

"The present is the key to the past."

From his belief, a doctrine of geologic origins was devised by simply projecting all natural processes back through time indefinitely. Though he acknowledged occasional local catastrophes, a major upheaval like the global Flood was strictly ruled out. **In fact Lyell actually wrote that his aim was to: "free the science from Moses."** [3] Does that sound like objective science, or a prejudice against the Bible?

CHARLES DARWIN

In our time the words "evolution" and "Darwin" are almost inseparable. At the age of 22 in 1831, with a degree in theology, Charles Darwin began a five-year world voyage as the naturalist aboard the British ship *Beagle*. He read Lyell's book and began to develop his concepts of biological origins according to the uniformitarian philosophy.

Darwin's dogmas of the evolution of species could never have gotten off the ground without Lyell's "gift" of millions of years. Thus, the foundation was laid for what came to be known as Darwinian evolution.

Charles Darwin didn't invent evolution. His grandfather, Erasmus, influenced his thinking. Others of the time could have received the credit, but it was Charles'

Some words from Darwin you likely haven't seen:

"I feel most deeply that this whole of Creation is too profound for human intellect. A dog might as well speculate on the mind of Newton! Let each man hope and believe what he can."

"To have the unthinking masses accept all that I say would be calamity...."

"What is true in my book will survive, and that which is error will be blown away as chaff." [7]

1859 book, *The Origin of Species*, which quickly exalted him to fame. He became the central figure of evolutionary thinking in a generation filled with revolution, humanist attitudes, and spiraling social upheaval.

GRADUALISM

"Gradualism" is a term often used regarding biological evolution in place of uniformitarianism, but the intended meaning is usually the same. Slow processes of genetic change, gradually going on generation after generation, through immense spans of time, are assumed to be the ultimate cause for virtually every living thing in nature. A modern zealot for Darwinism insists:

> Gradualness is of the essence. In the context of the fight against creationism, gradualism is more or less synonymous with evolution itself. **If you throw out gradualness you throw out the very thing that makes evolution more plausible than creation.** [4]

Working Definition of Evolution

> Evolution...a directional and essentially irreversible process occurring in time, which in its course gives rise to an increase of variety and an increasingly high level of organization in its products. Our present knowledge indeed forces us to the view that the whole of reality is **evolution, a single process of self-transformation.**

> Julian Huxley, famous evolutionist, atheist, and humanist of the 20th century [5]

In reality, as we'll see later, evolution is an anomaly to the law of biogenesis.

The Evolutionary Tree of Life

Most school children today are familiar with the typical diagram depicting how all forms of life supposedly evolved through the last two thousand million years. This chart is usually published with no explanation that it is purely a NATURALISTIC idea, based on the rejection of the Designer and His eyewitness account in Genesis. There is no evidence to substantiate it.

If there is no evidence, why believe the theory?

Many scientists have faced the reality that the subject of origins comes down to a personal and even dogmatic bias, as illustrated by the statement of evolutionist Richard Lewontin:

> We take the side of science [meaning evolution] in spite of the patent absurdity of some of its constructs… because we have a prior commitment… to materialism. It is not that the methods and institutions of science somehow compel us to accept a material explanation of the phenomenal world… we are forced by our *a priori* adherence to material causes… no matter how counterintuitive, how mystifying to the uninitiated… that materialism is absolute, for **we cannot allow a Divine Foot in the door.** [6]

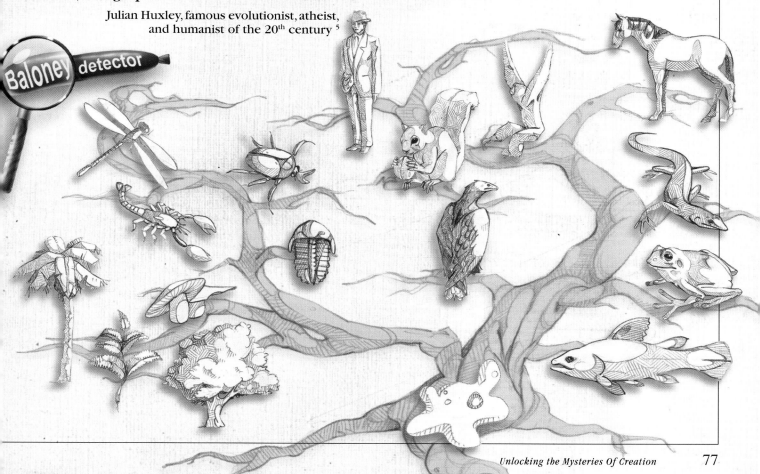

Can Good Science Embrace Bad Theory?

What Is Science?

The Oxford Dictionary defines science as "a branch of study...concerned either with a body of demonstrated truths or with observed facts systematically classified...under general laws, and which includes trustworthy methods for the discovery of new truth within its own domain."

G.G. Simpson, a leading evolutionist, has said: "It is inherent in any definition of science that **statements that cannot be checked by observation** are not really about anything...or at the very least, they **are not science.**"[8]

True Science Is
- Observable
- Demonstrable
- Repeatable

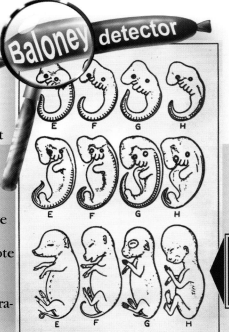

Does Science Contradict the Bible?

Some feel that God's Word and His world are separate and could conflict in apparent information. However, keep in mind that the Word of God speaks without apology about the origin and operation of the natural world. The straightforward narrative of creation was approved by Jesus Christ and by the apostles who wrote the New Testament. We must be careful with what we accept as science. True science will not contradict the Bible.

Science Mysteries

Despite all the work that has been done in all the science disciplines, there are major enduring problems. What is electricity? What is light? What is gravity?

Why do these phenomena behave so consistently? Some features of these things are known but not the basic "why" behind them.

Cautions for Science

Because there are so many disciplines of scientific study, being an authority in more than one realm is impossible for most men. Scientists are real people; they have the same human weaknesses and biases as non-scientists. Scientists who wish to be credible must be cautious not to make broad declarations on issues of religion or human spiritual matters in the name of science. When scientists do say things about origins, supernatural powers and spiritual things, you should be careful to examine and challenge them. Theories are not facts!

A Proven Hoax: Accepted for a Century as Science ...

- Embryonic recapitulation (the theory that human embryos develop by evolutionary stages of animal ancestry).

What Is a Theory?

The average person may suppose a theory is like a hunch or a guess, but scientists regard theories as accepted principles or concepts that spawn resea to explain natural phenomenon.

Have Scientists Been Proven Wrong Before?

False ideas have commonly been promoted by scientists of past generations only to be discarded by their descendants.

- The geocentric theory of astronomy.
- The idea that bloodletting is an appropriate medical procedure for reducing fever.
- Phlogiston was believed by 17th and 18th century scientists to be the magical ingredient of all combustible things that enabled them to burn. However, French chemist, Lavoisier, showed that oxygen was the key for fire (rapid oxidation) to burn.

Is Evolution Scientific?

No matter how you look at it, the theory of evolution must trace back to a point where inanimate matter became a living creature. Look at the recipe below for the foundation of this theory. Is it any wonder why many scientists like Dr. Henry Morris (who proposed the above idea) insist that **evolution does not even constitute a bona fide scientific theory!**

The Big Question

Bible believing Christians are often mocked by those who say: "Prove your eternal God created all this!"

But what's the alternative?

Why not ask the scorner:

"Prove to me that atoms are eternal! If God is not the Creator, how did the atoms get there?"

Does Change Equal Evolution?

Everyone who cares to study it can see that physiological changes sometime occur in living systems. Some call this "micro-evolution." We must acknowledge that the amount of change is always limited by existing genetic information within "kinds." The commonly held belief in evolution can be more specifically called "macro-evolution." It requires more than minuscule mutations to produce all the different living kinds. Indeed, TRANSMUTATION is the order of the philosophy, and that is the word that was used in the early 19th century before the word "evolution" became acceptable.

Mysteries of Evolution?

For any creation or upward development there are basically three things that are absolutely essential:

A PROGRAM to direct the growth.

A MECHANISM to energize the growth.

A PROTECTION SYSTEM to sustain the growth.

```
The Evolutionary Recipe.

1.UNKNOWN CHEMICALS
   in the primordial past... through...
2.UNKNOWN PROCESSES
   that no longer exist...produced...
3.UNKNOWN LIFE FORMS
   that are not to be found...
   but could, through
4.UNKNOWN REPRODUCTION METHODS,
   spawn new life...in an
5.UNKNOWN OCEANIC SOUP COMPLEX
   ...at an...
6.UNKNOWN TIME and PLACE
```

Fallacy #1 "It All Started with a Big Bang"

What Started It All?

According to materialistic evolutionary philosophers, the ultimate beginning of all life depends on a completely naturalistic premise that would understandably be thought of as destructive; namely, the big bang! They say the entire universe was once completely EMPTY…void… blank… just space…(in fact they insist that not even space existed) except for this one LONE "cosmic egg." The primeval ball of matter (which some say was essentially invisible) was just sitting there for aeons until one day it blew up with an unbelievable release of energy and heat.

Currently, it is common to believe that this mega-blast happened some 12 to 15 billion years ago. They believe the cosmic egg was an insignificant mass of matter but somehow it chanced to blow up, and things have been spinning off that explosion ever since.

From then on, with random processes, everything in the universe became increasingly complex. The sun and all the other stars eventually condensed out of swirling gas. The planets, including our Earth, also condensed out of dust particles hurling through space.

BUT WAIT! We need to ask an important question here.

Do Explosions Ever Increase Order?

TEST: If you take a stick of dynamite and go out and blow up a pile of lumber or other building material, is it going to end up more orderly? Perhaps it will look a bit more like a house after exploding!

But Some Materialists Protest: "Given a steady addition of energy, it's possible for order to come about. After all, the earth is an 'open system' to the sun's immense input of energy." The implication here is that the first primordial life could have popped up eventually on the earth with the provision of solar energy or whatever! [Though it seems strange to jump from a cosmic egg exploding, to the explosion of the first life in some warm pond on the earth, it is quite typical for both these unlikely events to be demanded in the evolutionary fantasy.]

HOLD IT! Another question must be asked here:

Does energy alone produce order?

Try another one:

How many lightning bolts does it take to bring a corpse to life?

Remember Frankenstein? Miracles like that only happen in the movies.

What Do Explosions Really Create?

When Mount St. Helens blew up in the state of Washington, U.S.A., we saw a good demonstration of what results from an explosion. It was a tremendously powerful blast, equal to many times the force of a nuclear bomb explosion. For miles around the mountain everything was desolated. In fact it was called a "dead zone."

Scientific articles repeatedly call the "big bang" a reasonable explanation for the origin of the universe. But there are physical realities in the heavens that defy the possibility of such an explosion ever causing the order we see. Big bang theorists have many insurmountable objections to their theory.

Some Insurmountable Objections:

1. The problem of missing mass. Dark matter (invisible matter in and outside of galaxies) should exist because galaxies appear to have stronger gravity fields than can be accounted for by their visible matter.[9]

2. Gas clouds dissipate outward (Boyle's Law).

3. Rarity of supernovas.

4. More stars dying than there are years to form them, and no observations of new stars forming.

5. Dependence on the theory that red shift means only expansion away from our observation while there are known alternative reasons for red shift.

6. Where's the big "hole" in the middle of the universe?

7. Spiral galaxies would not hold their shape for the time they are supposed to have existed.

ASTRONOMICAL "FINE TUNING?"

National Geographic magazine quoted one astronomer about the puzzling precision of the universe:

"To get a universe that has expanded as long as ours has (an assumption) without collapsing or having its matter coast away would have required extraordinary fine tuning. A Chicago physicist calculated that the odds of achieving that kind of precise expansion...would be the same as throwing an imaginary microscopic dart across the universe to the most distant quasar and hitting a bull's eye one millimeter in diameter."[10]

THINK ! Knowing the odds, why don't more astronomers confess that presently popular theories are totally inadequate to explain such astronomical precision?

EXTRAVAGANT IMPLAUSIBILITY?

To help grasp what current evolutionary thinkers propose about the "big bang," an article by one faithful writer may be helpful.[11]

Suppose you accept the big-bang theory.... Here's what you believe, roughly, according to the model proposed by Alan Guth, a physicist at the Massachusetts Institute of Technology.

You believe that, once upon a time, all the potential of the cosmos – all the potential for a firmament of 40 billion galaxies at last count – was packed into a point smaller than a proton. You believe that within this incipient cosmos was neither hypercompressed matter, nor super-dense energy, nor any tangible substance. It was a 'false vacuum' through which coursed a weight-less, empty quantum-mechanical probability framework called a 'scalar field.' You're probably not clear about what a scalar field is, but then neither are most PhDs.

Next, you believe that, when the big bang sounded, the universe expanded from a pinpoint to cosmological size in far less than one second – space hurtling outward in a torrent of pure physics, the bow wave of the new cosmos moving at trillions of times the speed of light.

Further, you believe that, as subatomic particles began to unbuckle from the inexplicable proto-reality, both matter and antimatter formed. Immediately, these commodities began to collide and annihilate themselves, vanishing as mysteriously as they came. The only reason our universe is here today is that the bang was slightly asymmetrical, its yield favoring matter over antimatter by about one part per 100 million. Because of this, when the stupendous cosmic commencement day ended, a residue of standard matter survived, and from it the galaxies formed. That is to say: You believe that a microscopic, transparent, empty point in primordial space-time contained not just one universe, but enough potential for 100 million universes.

It's wise to take the big bang hypothesis seriously [according to this secular author], since considerable evidence weighs in its favor. The galaxies are expanding away from one another as if they had once been in the same place, then hurled outward; the interstellar void is slightly warmer than absolute zero, suggesting the universe was once superheated by something much stronger than the output of stars; the earliest nebulae appear to be composed of precisely the mix of elements that big bang calculations suggest.

Yet, for sheer extravagant implausibility, nothing in theology or metaphysics can hold a candle to the Bang. Surely, if this description of the cosmic genesis came from the Bible or the Koran rather than the Massachusetts Institute of Technology, it would be treated as a preposterous myth.

Fallacy #2 "Time, the Magic Factor"

Evolutionary ideas require lots of time. Does 4+ billion years sound sufficient? Without the inconceivable "eons" of time, evolution hasn't got a chance!

Suppose you were a high school student assigned to write a term paper for your biology teacher. Imagine the look on his face if you proposed to elucidate on the theory that modern man evolved from single-celled amoebae in just two seconds. "Ridiculous," he would say, and rightfully so. But you are determined to come up with something new and scientifically stimulating so you think awhile and return with a new twist. "How about the idea that man evolved from amoebae in two billion years?" you ask. With a wry smile on his face, your teacher smugly tells you, "Now you've got the right idea."

Do you see what happened? Simply give it billions of years and:

the ridiculous becomes acceptable!

In fact, you've probably seen it stated somewhere by now that "given enough time, just about anything can happen!" That's one of the key fallacies of evolutionary thinking.

The late George Wald, professor of biology at Harvard University, rationalized that life is inevitable. He wrote:

Time is in fact the hero of the plot.
The time with which we have to deal is of the order of two billion years. What we regard as impossible on the basis of human experience is meaningless here. Given so much time, the 'impossible' becomes possible, the possible probable, and the probable virtually certain. One has only to wait: time itself performs the miracles. [12]

In the fairy tale, it was the kiss of a princess that turned the frog to a prince. Now we just wave the magic wand of MILLIONS OF YEARS and, presto...all life forms evolve on the earth.

But does evidence demand evolutionary time existed?

If you have studied statistics you know that the probability of an event happening does not become greater by allowing for more time. The probability is the same every day. The most probable things happen. Events with extremely low probability fall into the category of "improbable". **When a series of "improbable" events must occur in sequence at very specific times, the 'improbable' becomes impossible.**

Occasionally, a respected scientist has openly admitted the deficiency of any substantial proof for the vast time scales demanded by evolutionary philosophy. Astrophysicist John A. Eddy, solar astronomer at the High Altitude observatory at Boulder, Colorado is quoted with regard to the age of our sun:

There is no evidence based solely on solar observation, that the sun is 4.5 billion years old. I suspect that the sun is 4.5 billion years old. However, given some new and unexpected results to the contrary, and some time for frantic recalculation and theoretical readjustment, I suspect that we could live with Bishop Ussher's value for the age of the earth and sun. I don't think we have much in the way of observational evidence in astronomy to conflict with that. [Archbishop James Ussher lived in the late 16th and early 17th centuries. His careful biblical scholarship led him to calculate that the creation week described in Genesis actually took place in 4004 BC.][13]

What really happens as time marches on?

- Orderly things become disordered.
- New things get old and break down.
- Living things age and wear out.

This is called the second law of thermodynamics and it is one of the most constant laws of nature, found everywhere in the universe. It is sometimes referred to as the law of ENTROPY.

Which represents reality? Evolution or Entropy?

The Latin word from which "evolution" comes literally means "an out-rolling." It indicates something spiraling from the infinitesimal to all of reality. The word "entropy" comes from the Greek word meaning "in-turning."

Poles Apart

Evolution - change outward and upward

Entropy - change inward and downward

Every energy system in the universe wears out. Stars burn out, galaxies fly apart, energy becomes less and less useful through time.

Science is painfully aware of the rigidity of this natural law. The impossibility of the perpetual motion machine verifies the consistent observation that on the cosmic scale, **everything in our physical universe is running down.**

ENTROPY TIME

Thermodynamics and The Bible

The Two Laws Explained

FIRST LAW: Energy is conserved.

All existing processes of nature merely change energy from one form to another. In nature, energy is neither created nor destroyed. Matter itself (which is potential atomic energy) is maintained at a constant level. Processes change matter and energy from one form to others but the total quantity of energy in the universe always remains the same.

SECOND LAW: Energy dissipates.

As processes in nature occur, the total energy available is reduced to simpler forms with a consequent increase in what has been termed "entropy." As energy is used it becomes less available for further use. Part of the energy spent to produce something is always dissipated by radiation, friction, etc. It dissipates in space as non-recoverable heat. Ultimately, as things are going (and without God's intervention), the entire universe will end up filled with useless, stagnant, low-level heat energy.

Thermodynamics and Scripture

No discovery of true science takes God by surprise. We should expect that, in His wisdom, there would be hints of these truths found in His Word, the Bible.

1. The first law speaks of a total creation, originally completed, and now sustained by God's power.

2. The second law speaks of the curse of decay and death, brought on by man's sin, and causing an overall degeneration in everything.

The First Law in Scripture

The "conservation principle" of the first law of thermodynamics is easily understood in the context of a completed creation that is now being sustained by the Creator.

*"…All things **were created** through Him…" (past tense; notice it is not a continuing process).*

"…in Him all things consist" (Greek word for sustain; i.e. nothing is lost from it).
                              Colossians 1:16 & 17

"…He made…upholding all things by the word of His power."   *Hebrews 1:2 & 3*

"…by the word of God… the present heavens and earth… are reserved…kept in store…."
                              2 Peter 3:5 & 7

"…He commanded…created…established them forever."   *Psalm 148:5 & 6*

"…He…created…not one faileth."   *Isaiah 40:26*

"…Lord…made…all things…preservest them all."
                              Nehemiah 9:6

It is clear from Genesis chapter 1 and the first three verses of chapter 2 that whatever methods God used to create were stopped then. "He rested from all His work" and called it "very good." (Genesis 2:2 & 1:31)

The Second Law in Scripture

The "decay principle" is consistent with the entire Scripture in light of the curse in Eden.

"Even they will perish (the starry heavens), but Thou dost endure; and all of them will wear out like a garment...."   *Psalm 102:26*

As Romans 8:22 says: "the whole creation groans and suffers..." This seems contrary to God's original purposes for a creation that He called "very good." Death and decay was obviously not intended to prevail as it does now. Nor will it dominate the universe in the eternal kingdom as described in Revelation 22:3.

The result of man's sin described in Genesis 3:17 has even affected the ground, the very elements of the physical creation, including the dust of which Adam was made. Some aspect of the second law apparently began then. Before that, some higher law of perfection and preservation must have been in effect. Such supernatural care

was indeed in effect at least partially during Israel's 40-year episode in the wilderness (See Nehemiah 9:21).

The Bible makes clear the fact that the second law is not permanent. The Creator himself has promised that the curse of decay is not endless.

"...creation...will be set free from...the bondage of decay."   *Romans 8:21(RSV)*

"...no more curse...."   *Revelation 22:2*

THINK!

- **The second law proves that there had to be a beginning point once in time, or else all creation would be dead by now. The universe cannot have an infinite past.**

- **The first law proves that a Designer must have created all that exists, because no process in nature creates anything out of nothing.**

Fallacy #3 "Random Chance Produces All the Complexity of Living Things"

One of the most commonly heard beliefs of evolutionist dogma is the idea that everything in reality is only the result of pure chance. The infinite evidence in nature of orderliness, complexity and **design is not actually "real" to the evolutionist**, but instead is only "apparent." Because of the rejection of a supernatural intelligent designer, the implication of such thinking is that there really isn't any order in anything because **everything** is just the result of random (accidental) **CHANCE!**

THINK! Is it possible for the precise organization of parts in living systems to come into existence by mere chance?

What Is Random Chance?

When flipping a coin you have a chance of one in two that it will turn up heads. How likely is it that three objects would be arranged in a certain way if they were spilled out on the table? The chances can be figured by a simple mathematical equation. In this case it is expressed as "3!" and is called "three factorial." The calculation is figured this way:

$$1 \times 2 \times 3 = 6$$

Thus, there are six ways to arrange three parts in a simple sequence. When you start adding parts, the chances quickly grow very slim. For example, what are the chances of blindly arranging (by chance) 6 different items in one perfect order? The result is one out of 6! which is calculated as:

$$1 \times 2 \times 3 \times 4 \times 5 \times 6 = 720$$

For the time being, forget the problem of how the first **living cell** happened by chance, with its **multiplied millions of specially organized parts.** Let's look at a simple mathematical argument by examining the odds of arranging a simple system of just 200 parts by accident. How many ways can it be arranged? For example, the human skeleton has about 200 bones. Given only 200 fixed places to put them, what are the chances of randomly arranging those bones all in the right order on the first try?

The number is 200! (200 factorial), and it is immense. It is represented by the number 1 with 375 zeros after it. There is no way to imagine that number! Each time you try a new arrangement of your 200 parts you use up one of your "chances." Let's say you can systematically sort out your 200 parts a new way every second. Now remember, there is only one correct way out of 200! different possibilities.

How many seconds are there in a year?

   60 sec. x 60 min. = 3600 sec./hr.

   x 24 hours = 86,400 sec./day

   x 365 days = 31,536,000 sec./year

In a billion years?

   31,536,000,000,000,000 seconds

In ten billion years?

   315,360,000,000,000,000 seconds

How much is that?

Under a third of a billion "billions."

But 200! amounts to 1 with 375 zeros after it. The total number of electrons that could be packed into the entire universe would only total 1 with 130 zeros after it. Do you see how utterly ridiculous this gets? There isn't enough time to apply the chances needed to organize the designs in creation, even if you allow you allow billions of times the billions of years that evolutionists require.

But this is just one system of only 200 parts. What about the chances of all the millions of parts of the millions of organisms on earth coming together in certain ways with no design, no pattern, no system to plan and build them?

Before you even have a living organism you must have specific arrangements of protein molecules. Thousands of different proteins are required as building blocks of even the simplest bacteria. Humans have some 200,000 different proteins to make up all their organic structures. Most proteins are made of hundreds of amino acid molecules. Every amino acid must have at least one activating enzyme.

All these parts must be perfectly together at once just to provide for the construction of living organisms with all their diverse systems and organs. But you still don't have life. This is like having lots of paper, ink and lead type (as in an old fashioned printing shop) with nobody to write and assemble the sentences to make a book. **For something to work or to live** takes far more than raw materials or even organized parts.

Ask a mathematician what are the odds of accidentally discovering the winning ticket for the jackpot lottery lying on the ground by your trash can. Now ask what are the odds of finding ANOTHER winning ticket tomorrow in the same place. For even one intricately arranged genetic code (for something like an amoeba) to evolve by chance is similar to the chance of finding another winning lottery ticket **every day** for the next thousand years! In other words, you don't have to be a mathematician to conclude that such odds are utterly impossible. Yet, **this kind of trick** had to happen trillions of times over to produce **all** the systems of life on earth by chance.

It takes intelligence.

Oh Boy, another winning ticket!

If You Believe in Design by Chance, It's Not Because of the Evidence

When considering the subject of chance as it relates to evolution, consider one of the most complicated single things we know about: **the living cell.**

THINK! Each one of us came from one single fertilized cell. In the nucleus of that little dot, the master computer of what scientists call DNA contained the genetic programming for every aspect of the yet-undeveloped adult individual. Every organ, every nerve, every hair, even personality traits and behavioral patterns, as well as hair and skin color are programmed in those incredibly tiny chromosomes.

How Tiny Are These Mini-computers?

According to Ashley Montague in his book *Human Heredity,* the space occupied by all this data is incredibly small. If you could gather the programmed genetic coding for every human being on earth from his father's side alone it would only take up the space of less than half the size of an aspirin tablet! As impressive as microchips are to this computer generation of ours, this bio-micro-circuitry is beyond the incredible. But somehow, random chances had to be responsible if evolution is true.

It's been estimated that a single gene is between 4 and 50 millionths of an inch across. A half million genes would easily slide around in a hole made by an ordinary pinpoint!

Appreciating God's Amazing Design of Genes

Within the nucleus of every human cell, twenty-three pairs of chromosomes are intertwined like a wad of spaghetti in the nucleus. These are made of DNA. The DNA molecules contain all the thousands of genes which program living things to go on as they do. Genes mastermind the complexities of life in ways infinitely beyond man's understanding.

What Do Genes Do?

1. Determine all inherited traits like height and personality type.

2. Direct all growth processes like when and how baby teeth are pushed out by adult teeth.

3. Program all structural and system details for every organ in the body.

The Real Surprise

We now realize that each and every cell in your entire body **contains ALL the genetic coding for each of the other cells of your body!**

THINK! Somehow, despite each cell's vast sea of "blueprints," each cell always "knows" its own job and keeps it. A cell of the skin on the end of your nose has the genetic information to produce a fingernail cell... but it doesn't! Aren't you glad?

Dr. Michael Denton, a microbiologist without a commitment to conservative Christianity or biblical creation, challenged the evolutionary community in his important book, *Evolution: A Theory in Crisis*. He wrote:

> "**The complexity of the simplest known type of cell** is so great that it is impossible to accept that such an object could have been thrown together suddenly by some kind of freakish, vastly improbable event. Such an occurrence would be **indistinguishable from a miracle**."[14]

Is it any wonder why mathematicians can have serious doubts about evolution? Mathematician I. L. Cohen writes:

> "...any physical change of any size, shape or form is strictly the result of purposeful alignment of billions of nucleotides (in the DNA). Nature or species do not have the capacity to rearrange them nor to add to them... The only way we know for a DNA to be altered is through a meaningful intervention from an outside source of intelligence – one who knows what it is doing, such as our genetic engineers are now performing in the laboratories."[15]

Further in Mr. Cohen's article he writes:

> "In a certain sense, the debate transcends the confrontation between evolutionists and creationists. **We now have a debate within the scientific community itself; it is a confrontation between scientific objectivity and ingrained prejudice** – between logic and emotion – between fact and fiction." (pp. 6-7)

> "...In the final analysis, objective scientific logic has to prevail – no matter what the final result is – no matter how many time-honored idols have to be discarded in the process." (p. 8)

> "...After all, it is not the duty of science to defend the theory of evolution, and stick by it to the bitter end – no matter what illogical and unsupported conclusions it offers.... If in the process of impartial scientific logic, they find that creation by outside super-intelligence is the solution to our quandary, then let's cut the umbilical cord that tied us down to Darwin for such a long time. It is choking us and holding us back." (pp. 214-215)

> "... every single concept advanced by the theory of evolution (and amended thereafter) is imaginary as it is not supported by the scientifically established facts of microbiology, fossils, and mathematical probability concepts. Darwin was wrong." (p. 209)

> "...The theory of evolution may be the worst mistake made in science." (p. 210)

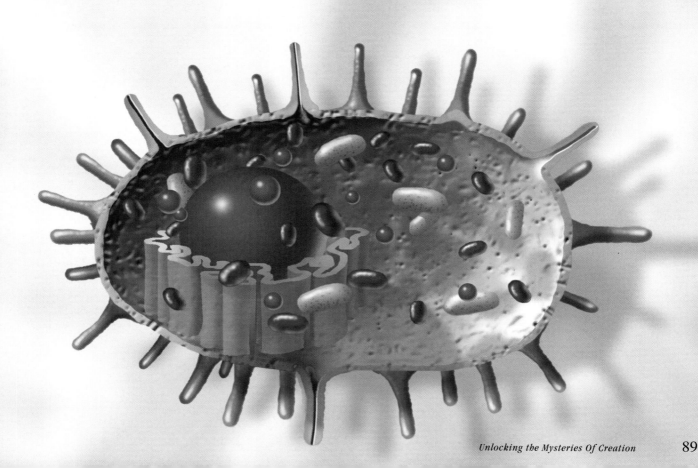

Fallacy #4 "Non-Living Matter Produced Life"

Spontaneous Generation

With random processes and time, according to materialistic thinking, some raw elements of the earth eventually became living organisms. This widely held belief is called spontaneous generation. Have you seen the colorful imaginary pictures used to impress young students with what many people believe is reality?

Medieval Science Revisited

In the Middle Ages virtually everybody in Europe **believed** that **rats and flies came from garbage,** and mice came from stored grain. After all, every town had a garbage dump on the outskirts and rats and flies were always seen emerging there. Farmers always found mice in their grain bins. **These were observed scientific facts!**

Finally an inquisitive researcher in the 17th century named Francesco Redi decided to test the popular misconception. In 1668, he placed some garbage in a sterile jar under a screen and just watched. What do you suppose happened? You guessed it. Nothing!

By the mid-1800s the famous scientist, Louis Pasteur, confirmed this reality with his discovery of microscopic "germs." That seems so obvious to us today. The conclusion? Rats and flies came from parent rats and flies! So a new scientific law came into popularity. It's called the law of...

Biogenesis

Bio means "life." **Genesis** means "beginning." Thus, biogenesis has come to be known as one of the **most universal laws of science**: life begins from life! It always works that way. Consequently, the theory of spontaneous generation was discarded as obsolete and scientifically unacceptable. Unacceptable, that is, until you reach the chapter on the modern "scientific" illusion of evolution in your high school biology text book.

THINK! Evolution doesn't even have the garbage! Evolution requires that life came from raw elements of the earth, and nothing more. With what scientists know today about life, how could spontaneous generation produce even a 'simple' life form or even the universal DNA code that is loaded with sophisticated programming? Can this kind of information be generated spontaneously?

What Does the Genesis Record Say?

"God said: 'Let the earth bring forth living creatures after their kind: cattle, and creeping things, and beasts of the earth after their kind;' and it was so" (Genesis 1:24).

God is the one supremely powerful source of all living creatures. He made them suddenly and completely, to reproduce after their kinds.

But wait! Hasn't life been "created" synthetically in a test tube? NO!

What Has Been Done in the Laboratory?

The famous Stanley Miller experiment, which is referenced in every evolutionary textbook, only produced amino acids, which are the building blocks of proteins.

Is it any wonder why some leading evolutionists have made open admissions of the failure of evolutionary theory to produce the evidence needed?

"..the macro-molecule-to-cell transition is a jump of fantastic dimensions, which lies beyond the range of testable hypothesis. In this area all is conjecture. The available facts do not provide a basis for postulating that cells arose on this planet."[17]

"After having chided the theologian for his reliance on myth and miracle, science found itself in the unenviable position of having to create a mythology of its own: namely, the assumption that what, after long effort could not be proved to take place today, had, in truth, taken place in the primeval past."[18]

"One has only to contemplate the magnitude of this task to **concede** that the **spontaneous generation** of a living organism is **impossible**. Yet we are here as a result, I believe, of spontaneous generation."[19]

Did you get that? A highly educated man (who is now dead) chose to believe what is known to be impossible rather than acknowledge the Creator. The words of the first chapter of Paul's letter to the Romans ring so loudly:

"Professing themselves to be wise, they became fools… And even as they did not like to retain God in their knowledge, God gave them over to a reprobate mind…" *Romans 1:22, 28*

Just think, as soon as I create life in a test tube, I'll disprove that absurd idea that it took intelligence to do it in the beginning... I'm such a genius!

They are a far cry from living cells. Worse yet, as noted by biologist Gary Parker, "the molecules Miller made did not include only the amino acids required for living systems; they included even greater quantities of amino acids that would be highly destructive to any 'evolving' life."[16]

THINK! Enzymes can only be produced by living cells, yet living cells absolutely require many specialized enzymes in order to survive!

The Complexity of the Single Cell

Not so long ago, the single living cell was called the "simple cell." But not any more! The bulk of the cell body was thought to be a jelly-like mass of what was called "protoplasm" (which simply means "living substance"). In 1963, Dr. George Palade, of the Rockefeller Institute in New York, discovered there is more to it than meets the eye. He found an amazingly intricate system throughout the protoplasm. Today it's called the "endoplasmic reticulum"

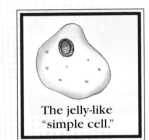

The jelly-like "simple cell."

(E.R. for short). This vast labyrinth of incredibly fine tubes and chains of minute bags totally permeates the entire cell body. The E.R. has been described as one of the most complicated and beautiful structures in the universe. Another long-held concept of modern science has crumbled. The idea that molecules just bang around haphazardly in the jelly-like protoplasm has been discarded in light of new discoveries.

Scientists today tell us that the single cell is more complex than a large city.

As in a city, so it is in a cell. Systems are working and jobs are being done by the thousands. But unlike cities, the cell functions perfectly with no breakdowns!

The adult human can have as many as 100 trillion of these walled cities. Think of the systems working constantly:

- design structures
- energy generators
- invasion guards
- transport systems
- food factories
- protective barriers
- waste disposal systems
- communication links within and outside the cell city

All these functions must work in harmony with each other and with the surrounding cells via communication links to maintain the health and fundamental activity of the cell. Typical cells contain thousands of specific proteins that are vital for life. Each protein is a chain made from up to 1,000 amino acid molecules in specific sequence. No cell could ever operate without exact parts that had to exist and function together from the start. Even the parts are too complex to function without intelligent plans.

Some Amazing Insights

THINK! How important is the "skin" of a living cell?

The surface membrane of a cell is amazingly delicate but very "powerful" in its control of the cell system. It's less than a third of a millionth of an inch thick! It controls the entry and exit of everything for the cell. It behaves as if it had the senses of taste and smell. When a desirable molecule floats by, the membrane forms a little "finger" that reaches out and pulls the needed nutrient inside. Vital chemical enzymes coat the skin of the cell, transferring information to and from other cells. The cell could not survive without these enzymes functioning precisely as they always do.

Molecular Machines That Manufacture Their Own Replacement Parts

So here is a microscopic "city" filled with super tiny machines that replicate themselves automatically. Thousands of different protein compounds are needed in every cell to do this. All of them are made of chains of amino acid compounds, mathematically precise.

The RNA molecule is the messenger in the metropolis of a cell. It looks like DNA, but has a passport to leave the nucleus. With incredible speed the RNA molecule acts like a computer printer. Let's see what happens.

1st The master DNA and the messenger RNA intertwine in a split second.

2nd The DNA instantly imprints a section of its code on the RNA, and then separates from it.

3rd The RNA rushes to the edge of the cell city to transfer its code to enzymes, making copies one after another in rapid-fire succession.

4th Each enzyme is commissioned by this code to do a particular job somewhere in the larger organism.

Another Mystery

The whole mass of cells within a body communicates by RNA. According to experts, they "somehow cooperate" to act like a dog, a fish, a man, or whatever the organism is supposed to be. That "somehow" is the insoluble mystery to secular scientists who choose not to begin their research on the factually solid foundation of God and His Word.

"O the depth of the riches both of the wisdom and knowledge of God! How unsearchable are His judgments and His ways past finding out!"

Romans 11:33

A Curious Note

Scientists have discovered that DNA is found in the nucleus of ALL living cells with the exception of red blood cells and a few certain viruses. Isn't it strange that the one component that science now calls the "mysterious basis of ALL life" (DNA) is NOT in the place God's Word tells us where life is found?

"The life of the flesh is in the blood...."

Leviticus 17:11

"How precious also are thy thoughts unto me, O God! How vast is the sum of them!"

Psalm 139:17

"It is astonishing to think that this remarkable piece of machinery, which possesses the ultimate capacity to construct every living thing that ever existed on earth, from giant redwood trees to the human brain, can construct all its own components in a matter of minutes and weigh less than 10^{-16} grams. It is of the order of several thousand million million times smaller than the smallest piece of functional machinery ever constructed by man" (until man invented nanotechnology).[20]

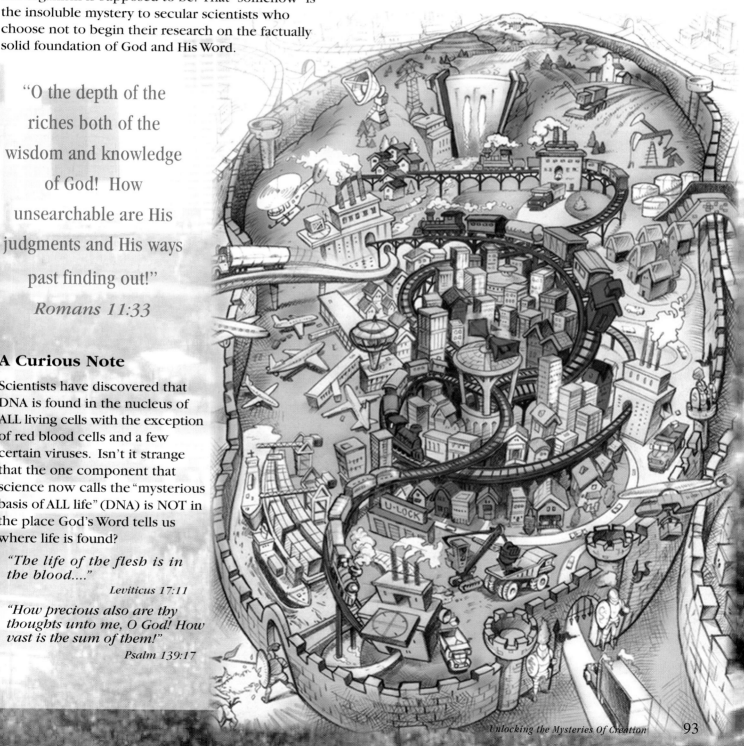

Inventions by Design
or Flukes of Mother Nature?

Have you heard the term BIONICS?

Bionics is a modern branch of electronic engineering specializing in inventing machines that imitate natural features of living creatures.

Keep in mind what evolution requires. Blind chance and some mysterious self-organizing phenomena of matter are alone responsible for all design in living creatures.

Sonar is a bionic invention.

THINK! Where did the idea of sonar originate? Who had it first?

The curious little bat had sonar (or echo-location) first! Echo-location is one of the most complex functions found in animals. It works like sonar in a submarine. Nobody would dare say that modern sonar technology just magically happened "by chance." Yet how is it that so many so-called intellectuals insist that no Designer planned the intricate sonar system of the bat? Is it reasonable that random accidents never witnessed by anyone brought about all that complexity?

To use echo-location, an animal needs a voice box capable of producing very high pitched sounds, plus ears that can hear very high pitched sounds, connected to a brain that can interpret the reflected sounds in 3D, and build a picture of the environment from the sound waves. Like sonar in a submarine, echoloca-tion is useless until all components are completely functional and interconnected. Another sonar, that of the dolphin, is said to be a thousand times more sophisticated than the sonar technology of our modern nuclear submarines. So it requires a lot of faith to believe it evolved even once, let alone twice.

What is the bionic equivalent to the human system of remembered sight? The video camera and recorder, genius invention that it is, can only be called "crude" when compared to the amazing complexity and sensitivity of the **living** technicolor visual system of the human eye and computer brain.

Darwin's amazement at the eye caused him to write in his first book:

"**To suppose that the eye**, with all its inimitable contrivances for adjusting the focus to different distances, for admitting different amounts of light, and for the correction of spherical and chromatic aberration, **could have been formed by natural selection,** seems, I freely confess, **absurd in the highest possible degree...**"[21]

If you notice that something is "absurd," wouldn't you be inclined to find a more reasonable explanation?

Designed things had a Designer.

Let's say you were out taking a stroll one day along the lakeshore. You happened to look down and spot a piece of flint that catches your attention. You bend over to pick it up...brush off the sand...and...lo and behold, you realize you've discovered an ancient Indian arrowhead. You marvel at its careful design; a simple bit of technology yet profoundly informative. What does it tell you?

SOMEONE MADE THIS! You realize that fact instinctively. There is no doubt about it. The symmetry of the little point gives it away. It is not the product of random erosion over the years as the waves of the lake lapped its shores. No! A thinking person designed and formed this little tool. (This event actually occurred when I was in graduate school, studying for the museum profession).

Now, turn to the wild flower just a step away. There's a Monarch butterfly hovering over its bright petals. You ponder the design, the complexity, and the detailed intricacy of that little plant-animal relationship. What a wonder you are experiencing. And did you ever wonder why God made the different kinds of flowers to bloom at different times? One reason is so the bees, butterflies and insects that live from flowers have a continuing supply of food.

"Who knoweth not in all these that the hand of the Lord hath wrought this?   *Job 12:9*

The idea of computers, or any other relic of technology, just happening by sheer chance, (not to mention the astounding complexity of the human brain) has about as much chance as a monkey typing the unabridged dictionary perfectly by chance (i.e. NONE).

Sir Fred Hoyle, the well-known English astronomer and professor at Cambridge University admitted: "The chance that higher life forms might have emerged in this way is comparable with the chance that a tornado sweeping through a junkyard might assemble a Boeing 747 from the materials therein."[22]

Fallacy #5 "Simple Forms Develop into Complex Forms of Life in Time"

According to evolutionary thinking, life began on earth two or three billion years ago and proceeded through thousands of generations to develop from so-called "simple" forms into more "complex" forms. The earliest forms of life are often called "primitive," or "lower forms."

Beware! When you read or hear evolution-biased accounts, the words "primitive" and "simple" are deliberately used to convey the idea that evolution has really happened.

THINK! When you realize the incredible design and complexity of even the so-called "simple cell," it becomes very clear that **NO LIFE IS SIMPLE!**

Questions:

- Is life simpler because it is smaller?
- Are larger forms more complex than smaller on
- Is there such a thing as "primitive" fish, frog or mammal? Or are they all really just as advanced any other?
- What defines primitive? Does the definition im a gradual progression of evolution?

How Did This Supposed Developmer Happen?

Mutation and natural selection are the assumed miracle-workers of this long evolutionary ch: of events.

The famous evolutionist and biologist, Ernst Mayr wrote:

> "It must not be forgotten that mutation is the ultimate source of all genetic variation found in natural populations and the only new material available for natural selection to work on." [23]

THINK! Aren't mutants deformed or distort in some way? If someone called you a mutan you wouldn't exactly take it as a compliment. That's because we all know what mutants are. Why do many seem to forget this when theorizi on evolution?

What Is a Mutation?

A mutation is a **mistaken change** in the genetic code, a loss of information.

- Most are considered neutral; they cause no noticeable change in the living creature.

- Harmful mutations are **rare.**

- Ultimately beneficial mutations **have not been verified.**

This mutant cherry tomato looks like it grew ears.... Obviously not a normal tomato!

What Have Mutation Experiments Proven?

Fruit flies have been subjected to all manner of mutant-producing experiments in the laboratory. In one experiment, these little bugs were bombarded with radiation and put through the torture test for

over 1,500 generations to synthetically produce mutant offspring.

The Result?

- Distorted, damaged fruit flies were produced plentifully.

- Shriveled wings – crooked bodies.

- Weak eyes, no eyes – sterile flies.

But never once has an improved fruit fly been spawned; much less a new kind of creature!

Respected evolutionist, Dr. Pierre-Paul Grassé concluded, "No matter how numerous they may be, mutations do not produce any kind of evolution."[24]

Though public perception may naively believe that mutations produced improved species, evolutionists who know better have often admitted they are generally useless, detrimental or deadly. [25]

When you understand the nature of mutations, you don't have to be a Nobel scientist to realize that: "Trying to improve an organism by mutation is like trying to improve a Swiss watch by dropping it and bending one of its wheels. Improving life by random mutation has a probability of zero."[26]

Mutated fruit flies have never proven to be more beneficial to the ultimate success of the fruit fly.

What about Natural Selection?

Is natural selection really the major cause of evolution as explained in popular textbooks?

THINK ! Natural selection, in reality, is God's natural way of picking the best-suited and strongest individuals of a given kind in order to perpetuate the furtherance of that kind.

Inferior oddities are automatically weeded out by natural selection. The best is preserved to maintain genetic stability. What does this tell you would happen to mutants?

The Classic Example: The Peppered Moth

Many school texts and nature books with a pro-evolution bias have referred to the peppered moth of England as a good example of "evolution in action."

Students are impressed to believe that the light variety of *Biston betularia* (the peppered moth) evolved into the dark variety during the industrial revolution as a result of factory soot coating the tree trunks where the moths landed.

What Really Happened?

Actually, it is a known fact that both light and dark varieties of the peppered moth lived in England prior to the industrial revolution. The dark ones were rare. It was supposed that as the light colored trees darkened with soot, the dominant white moths landing on the trees apparently became more visible to predatory birds. The dark moths supposedly were now better camouflaged so their numbers surpassed the white ones. Is that evolution? **No**, of course not, because what was there before was there after, just in different amounts. But why is it deceptively used as a "proof" of evolution?

Darwin Rejected Natural Selection

Even Darwin abandoned the concept of natural selection as an explanation for evolution. In the sixth edition of his famous book he wrote:

> Natural selection is incompetent to account for the incipient stages of useful structures....[28]

What about Dogs?

Are dogs an example of evolution? There are at least 200 modern breeds of dog today. They all trace their ancestry to a few original dog stocks several hundred years ago or more. Actually, this is an example of breeding selectively. It's the same genetic process responsible for the hybridization of cattle or roses. From the tiny Chihuahua to the great St. Bernard, they are all still dogs. Left to themselves for several generations, you'll end up with mongrels. But never once has some offspring turned up with a curious "meow," or a pig snout, or some other totally new and beneficial feature not found in the dog genetic pool.

"God gives it a body as it has pleased Him, and to every seed his own body. All flesh is not the same flesh: but there is one kind of flesh of men, another flesh of beasts, another of fishes, and another of birds."

1 Corinthians 15:38 & 39

Another Fraud Exposed?

Dr. Kettlewell of Oxford University in England introduced the peppered moth melanism story in 1955 as evidence for evolution by natural selection. Since then it has appeared in almost every biology text as evidence for evolution. Yet there are serious problems with Kettlewell's story:[27]

1) The moth does not rest on tree trunks! Exactly two moths have been seen on tree trunks in more than 40 years. Kettlewell actually GLUED two dead moths on the tree trunk to take his famous photograph that was published in so many school books.

2) Moths have no tendency to choose matching backgrounds.

3) Kettlewell's results have not been replicated in later studies.

4) The shift in moth population did occur, but took place well before new lichens grew on polluted trees. A parallel shift in moth population occurred in U.S. industrial areas but there was no change in lichens.

The peppered moth is just another example of a so-called proof for evolution that was widely published for many years and then quietly deleted when the truth was finally researched by other scientists. Perhaps the "expert's" pattern of "straining at gnats" has led many to "swallow a camel" of deception. Beware, the next time you hear about some sensationalized "proof" for evolution in action.

Is Life on Earth Actually Getting Better? More Complex?

When you look at the evidence for the history of life on earth, what do you find? Is life on earth really getting better or more complex, as evolutionist's ideas insist? Actually, DEGENERATION and EXTINCTION are the rule! The evidence is well documented and you'll be able to add many good examples to the list presented here.

The Chambered Nautilus

Source: *National Geographic*, January 1976

This fascinating shellfish is explored in the *Geographic*, telling of their supposed "progress over the years."

Quoting from the article:

"It remains essentially the same as its ancestors of 180 million years ago...a living link with the past."

Notice this:

"Some 3,500 nautiloid species once flourished. A nine-foot fossil turned up recently in Arkansas." Now, "...fewer than half a dozen species still exist...time has whittled these descendants to about eight inches."

HOLD IT! Did they say **"progress?"** But it's **"essentially the same."** Is progress defined by **"whittling"** the species from 3,500 to only 6, and cutting down their size from nine feet to just eight inches? Does this evidence convince us of evolution, or does it fit more closely with what we should expect if the creation of distinct kinds is true?

Is a Bat Always a Bat?

The picture of a fossilized bat skeleton has been published in a number of places. Experts regard it as the oldest bat remains known, and it supposedly is 50 million years old! (dated by the hypothetical chart of geologic history). The amazing mystery is revealed in the fact that the details of the skeleton are virtually the same as those of modern bats!

Ammonite, extinct Cephalopod: close relative to the Chambered Nautilus

Charles Darwin admitted in his book, *Origin of Species*: "If it could be demonstrated that any complex organ existed which could not possibly have been formed by numerous, successive, slight modifications, my theory would absolutely break down."

Origin of Species, 6th edition (1988), New York Univ. Press, New York, p. 154

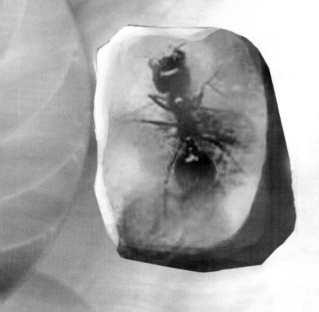

What about Insects?

Source: *National Geographic*, January 1981

A fossilized cockroach is pictured in this article about the amazing bug **"designed for survival."** According to *Geographic*, the "300-million-year-old fossil imprint...shows that roaches have changed but little since their world debut more than 320 million years ago."

THINK! If they are the same today as they've been for that long, stop to think what "their world debut" must have been like when they first came on stage! If roaches are the same as they ever were, so far as the evidence indicates, then why didn't the bug evolve into some "more complex" form? Yet this is **exactly the evidence to be expected if creation was divinely planned** and creatures really do reproduce "after their kind."

What Was Life Like in the Ancient Tropical World?

Source: *National Geographic*, September 1977

In a variety of places on Earth, ancient insects have been found encased in petrified tree sap. This **amber** has a special story of its own that should give us understanding on how things have evolved over the aeons.

A beautifully preserved ant is declared to be **100 million years old.** It's still an ant! And the remarkable thing is that it resembles types of ants living today.

The praying mantis in amber is fully like its descendants living today, yet it's dated by the evolutionary scale at 40 million years old. And imagine if you can, how an ordinary housefly and termite managed not to change into something more desirable in all the "millions of years" since they were stuck in that sap.

Keep Exploring

Some writers have called this phenomenon, "fixity of kinds." Although feature variations have occurred within kinds through time, there are distinct boundaries. That's why we do not find one kind of creature shading gradually into another kind.

The earliest fossil examples have the same body design as living examples. The excellent documentary video by Creation Research of Australia with John Mackay is an excellent source for you to discover more "Living Fossils."

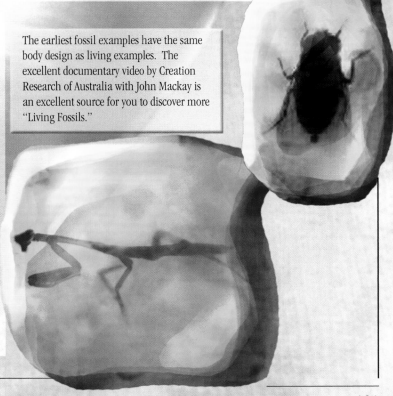

Do Fossils Show a Gradual Transition of Evolving Animal Kinds?

Have Fossil Links Been Found?

If evolution were true, links between kinds should be found plentifully. Paleontologists have been digging fossilized remains of animals out of the earth for years now, but what have they found to prove evolution? Darwin himself predicted that abundant fossilized links would be found showing transitions from one kind of creature to another.

Think ! **If evolution were true, why don't we see all manner of living transitional kinds?**

Has anyone ever seen a fish emerge from the sea to become an amphibian? You may have seen drawings of what **"must have been"** between fish and frogs but has such a creature ever really been seen?

NO! And in case someone tells you the oriental mudskipper is supposed proof, help them keep in mind that it is still a 100 percent fish!

Whales with Legs?

The evolution of whales has been quite a mystery that has evoked some rather imaginative scenarios. Supposedly, a land-dwelling mammal somehow gradually made it back to the sea, gave up his legs for a spout and a few other things, and in time, the sea monsters of all time just "evolved."

One recent article from a popular science magazine reported the conclusion that one paleontologist has drawn from his recent "find."[29]

According to Philip Gingerich, whales may have walked on four legs 50 million years ago. He found a skull and several teeth and came to the conclusion that they belonged to an ancient walking whale. Watch the article draw some interesting associations.

"How did he know that the piece of skull and teeth had anything to do with a whale, let alone a walking whale?

"Clue: the sediment in which the skull was found had been either a seashore or a riverbank. Conclusion: the creature lived in or near the water. Clue: the teeth were almost identical with those from known primitive whale fossils found on the west coast of India. Conclusion: the creature was probably a whale."

Watch carefully as the "scientific" conclusion to the article is drawn.

"It must have been heading in the direction of modern whales but wasn't quite there. So its legs probably had not yet evolved into flippers. Most of all though he (Gingerich) hopes to find leg bones belonging to the whale species. 'It is possible that we will find some,' he says, 'but **we will be lucky if we do.**'"

You may draw your own conclusions about that logic.

I gotta evolve some fins.

The Supposed Evolution of Birds

The remains of an ancient bird, caught in flooded silt and preserved, have been called *Archaeopteryx*. For years it has been proposed as a link between reptiles and birds.

Why would some believe this is a link when it is fully feathered and with a bone structure that can be defined as true bird?

As Dr. Duane Gish points out in *Evolution: The Fossils Say No!*, the primary reason some make a correlation between reptiles and archaeopteryx is because of its clawed wingtips and a beak full of teeth. Dr. Gish brilliantly points out the truth about this fossil. Living birds today have claws on their wingtips and they are 100 percent bird. Teeth in an ancient bird say nothing of its ancestral connection to lizards. Now there is another problem. Since 1984, bird fossil discoveries in Texas have been "dated" by evolutionists to be millions of years older than the dates assigned to archaeopteryx! And these are fully modern birds.

This model of a bird with a lizard-like head is similar to how archaeopteryx has been portrayed in museums everywhere. The fossil impression of the skeleton and feathers fills the background of this page.

So How Did Birds Evolve?

Some evolutionists believe that birds evolved their ability to fly by jumping out of trees.

According to a feature article in the *1980 Science Yearbook*:

> "**Paleontologists assumed that the bird's ancestors** [which have never been found] **learned to climb trees to escape from predators and to seek insect food. Once the 'bird' was in a tree, feathers and wings evolved** [as if by magic] **to aid in gliding from branch to branch.**"

Is this popular perception really how dinosaurs sprouted wings?

Others Have Different Ideas

According to some "experts," the primitive grandmother of all birds...

> "ran along the ground chasing flying insects which they nabbed with their teeth or front legs! Longer feathers on the front legs then evolved to act as an insect 'net,' and so the legs simply became wings. Then they used the wings to make flapping leaps after insects."

As the report concludes: "**If archaeopteryx did not fly, it was just on the verge.**" But it forgot to mention that all this baloney has no evidence to support it.

> **Controversial investigations by British evolutionist, Sir Fred Hoyle, now cast suspicion that archaeopteryx was fraudulently fabricated by its "discoverers" in the mid-1800s. Even though we have zero evidence for the evolution of birds, some popularizers and museums today are saying that modern birds are really feathered dinosaurs... highly evolved of course!**

Some Famous Fossil "Connections" To Evolution

There are many places on earth where fish fossils have been found. For decades the fish called "coelacanth" (pronounced SEE-luh-kanth) was called a "transitional form." It was supposed to be an evolutionary "link" between fish and amphibian. The coelacanth was presumed to have been extinct for the last 70 million years. So, it was used as an "index fossil" to assign ages to geologic layers. If you chanced upon one of their petrified skeletons you "knew" you were into some 70 million-year-old rock.

A Surprising Mystery

In 1938, fishermen hauled in a live coelacanth off the coast of Madagascar. It was picked up by an alert woman in the local fish market and was submitted to a scientist. He properly identified it as a genuine coelacanth. Since then, over thirty specimens have been caught alive. Reports of coelacanth are common in literature dealing with fossils. It is often featured as a "living link" with the past, but rarely are people made aware of the important implications of this so-called "primitive" living fish.

The coelacanth is obviously not an evolutionary link at all! It did not become extinct 70 million years ago. If you discovered a fossil coelacanth it would tell you nothing of the age of the rock in which you found it. Again, the evidence supports the concept that kinds remain essentially the same. There is no evolution of kinds from other kinds.

It's a coelocanth! This rock must be 70,000,000 years old!

The Evolution Of Horses: Fact Or Fiction?

The so-called "horse series" of transitional forms has been widely taught and published. Biology teachers have been given this convincing looking chart to demonstrate a clear example of evolution to their classes. What are the facts?

The creatures shown in the chart are based on bones found in widely diverse locations: India, North America, South America, and Europe. They've been cleverly arranged in the chart with the small ones at the bottom and the large ones at the top. This does indeed make the order look like a progressive evolution, doesn't it?

Some specimens were questionable. Some scientists felt they should be included as links to hippopotamus or some other animal, rather than to the modern horse.

In some places on earth, two or more of the horse series have been found buried together in the same layer of rock. This indicates they lived at the same time and were not distantly related transitions. In Florida the fossil bones of seven different kinds of horse have been found together in the same cataclysmic burial place. In one South American location the order of the chart is apparently upside down. The modern horse is deeper, and the more "primitive" three-toed type is higher up in the rock strata.

Another mystery seldom told is that the specimens in the chart have an unusually illogical evolution of the rib cage. Starting at the bottom, the smallest one has 15 pairs of ribs. The next one up has 18. Then comes a "higher" horse with 19, but then it drops down to 18 pairs at the top. Is evolution displayed in this?

It's rock hard evidence like this that leads many scientists to recognize the difficulty of accepting evolution from the record of the past. Dr. David Raup, curator of the Field Museum of Natural History in Chicago, wrote: "Classic cases...such as evolution of the horse in North America have had to be modified or discarded as the result of more detailed information."

Puzzling To Darwin

In his book, *The Origin of Species* (1859), Darwin admitted:

"As by this theory innumerable **transitional forms must have existed, why do we not find** them embedded in countless numbers in the crust of the earth? The number of intermediate links between all living and extinct species must have been inconceivably great! Geology assuredly does not reveal any such finely graduated organic chain; and this, perhaps, **is the most obvious and gravest objection which can be urged against my theory.** The explanation lies, as I believe, in the extreme imperfection of the geological record."[32]

As *Newsweek* magazine pointed out (11/3/80): "**Missing links are the rule.** The more scientists have searched for the transitional forms between species, the more they have become frustrated."

Another expert wrote: "The known fossil record fails to document a single example of phyletic evolution accomplishing a major morphologic transition and hence offers no evidence that the gradualistic model can be valid"[33] (Don't let the big words scare you).

Stephen Gould of Harvard tells it like it is, writing: **"The family trees that adorn our textbooks have data only at the tips and nodes of their branches; the rest is inference, however reasonable, not the evidence of fossils."**[34]

Gould goes further to say: **"The absence of fossil evidence for intermediary stages between major transitions in organic design, indeed our inability, even in our imagination, to construct functional intermediates in many cases, has been a persistent and nagging problem for gradualistic accounts of evolution."**[35]

THINK! Many students have been tricked into believing that the fossil record proves evolution. Now WE KNOW that fossils do not supply the evidence that evolutionists hoped for.

Does the Geologic Chart Prove Evolution?

What Is the Geologic Chart?

It was first devised by creationists in the late 1700s to chart relationships of rock strata formed before, during and after the Flood. Now this common chart of earth's geologic history is used to support the "theory" of evolution. It diagrams over a half billion supposed years of mostly sedimentary rock deposits in which are buried the fossilized remains of creatures on the path of evolution.

No such complete progression of rock layers and corresponding fossils is found anywhere on planet Earth!

How Are Fossils Dated?

When fossils are found and displayed, the dates you hear about are based on this chart. It doesn't matter if the fossil is a clam, fish, or dinosaur, the age for the specimen is determined by the chart.

Let's see how it works

If you find fossil trilobites and want to know their age you naturally take them to the paleontologist at a local university. He'll ask where you found them. Then he may identify the layer as "Cambrian." Next he'll take you to the geologic chart on the wall. "Here it is on the chart," he'll say. "So how old are the fossils?", you ask. He replies, "About 450 million years."

Being naively curious you ask, "How do you know that?" He replies, sounding slightly intimidating, "Because they're in the Cambrian layer, of course."

Now you're challenged to get to the heart of all this, so you walk across the hall to see the geology professor. You show him a picture of the place you found your fossil trilobites. You don't want to confuse the issue with all you've just learned, so you simply ask him, "Sir, how old would you judge this layer of rock to be?"

"That's easy," he says. "What kind of fossils have you found in it?"

"Trilobites," you confidently reply.

"Easier still," he says. "It's about 450 million years old."

Wanting to be doubly sure of the facts, you inquire, "How did you know that?"

Possibly a bit indignant at your revealing investigation, he tells you, "If trilobites are there, then we know the layer is that old."

ERA	PERIOD		
CENOZOIC 2,500,000 Yrs / 66,000,000 Yrs	QUATERNARY / TERTIARY		AGE of MAMMALS / MODERN SEED PLANTS
MESOZOIC 144,000,000 Yrs / 208,000,000 Yrs / 245,000,000 Yrs	CRETACEOUS / JURASSIC / TRIASSIC		AGE of AMMONITES and REPTILES / ANCIENT SEED PLANTS
PALAEOZOIC 286,000,000 Yrs / 360,000,000 Yrs / 408,000,000 Yrs / 438,000,000 Yrs / 505,000,000 Yrs / 570,000,000 Yrs	PERMIAN / CARBONIFEROUS / DEVONIAN / SILURIAN / ORDOVICIAN / CAMBRIAN		AGE of AMPHIBIANS / AGE of FISHES and CORALS / AGE of INVERTEBRATES / SPORE BEARING PLANTS / SEA-WEEDS
PROTEROZOIC 700,000,000 Yrs / **ARCHAEOZOIC** 4,600,000,000 Yrs	PRECAMBRIAN		RISE of INVERTEBRATES

SHOWING DEVELOPMENT OF ROCKS — DIVISIONS OF GEOLOGICAL TIME PLANT AND ANIMAL LIFE — DOMINANT LIFE

Cambridge University's Geologist, R.R. Rastall, wrote in *Encyclopaedia Britannica* (1956, vol. 10, p. 168): "...geologists are arguing in a circle. The succession of organisms has been determined by the study of their remains imbedded in the rocks, and the relative ages of the rocks are determined by the remains of organisms they contain."[36]

Keep in mind that there is no radiometric method (or any chemical process) for the direct absolute dating of fossils. Fossils are dated by the rocks and the rocks are dated by the fossils. As one evolutionist writes: "I can think of no cases of radioactive decay being used to date fossils.... Ever since William Smith at the beginning of the 19th century, fossils have been and still are the best and most accurate method of dating and correlating the rocks in which they occur."[37]

Living Fossils?

A little sea creature named Apus was called a "living fossil" because it is so similar to the trilobite. Why didn't these creatures phase out altogether as evolution marched on to greater things? Some feel that since trilobites lived on the ocean floors it's entirely possible to also find living trilobites. There goes another "index fossil" if they are found!

What else has been learned at the bottom of the chart?

New studies were reported about the Burgess shale beds at Field, B.C., known famously for the trilobite fossils found there. But there is a vast deposit of many other complex animals found there as well. Quoting from the article, "Animals...are so bizarre in this massive graveyard that not even the most imaginative paleontologist has been able to

connect them to any family of modern animals."[38] And note the "expert" conclusion: **"The textbook evolutionary tree,** with everything traced back to a few common ancestors, **is inaccurate!"**

Major Mystery of Evolution?

Besides trilobites, it has been repeatedly confirmed that the "Cambrian" layer is loaded with every major invertebrate form of life including sponges, corals, worms, jellyfish, crustaceans, and mollusks. Billions of fossils of well-developed marine organisms are found in this "basement layer of life" worldwide. Below that layer, in what has been called Pre-Cambrian, there is **a striking absence of multi-cellular life forms.**

Geologist H.S. Ladd, in the *Geological Society of America Memoir* wrote: "Most paleontologists...[are] ignoring the **most important missing link of all.** Indeed the missing Pre-Cambrian record cannot properly be described as a link for it is in reality, about nine-tenths of the chain of life: the first nine-tenths!"[39]

It's due to this problem that the famous evolutionist George Gaylord Simpson said that the absence of Pre-Cambrian fossils is **"the major mystery of the history of life."**[40]

Earth's fossils show exactly what you would expect if the creation explanation is true. Multiplied living kinds appear abruptly with no "primitive" links leading up to them.

THINK!
Does this mean that typical museum displays and textbook charts showing the familiar molecule-to-man idea are misleading?

Evolution's New Realms of Reason

With all these "mysteries" to cloud the issue of evolution, some expert evolution proponents have escaped to new realms of "reason."

The Theory of Panspermia

Dr. Francis Crick, biologist and Nobel Prize winner, famous for his co-discovery of DNA, wrote:

> An honest man, armed with all the knowledge available to us now, could only state that in some sense, the origin of life appears at the moment to be almost a miracle, so many are the conditions which would have had to have been satisfied to get it going. [41]

Yet, he is foremost among some, who propose a "superman slant," insisting "...that life on earth may have sprung from tiny organisms from a distant planet, sent here by space ship as part of a deliberate act of seeding." [42]

Hey Hun, what about that barren planet? Let's start life there.

But he fails to say anything about life being "planted" here by God!

Many Scientists Are in a Quandary about Darwin's Theory

Newsweek magazine (4/8/85) carried an article titled "Science Contra Darwin." Unknown to much of the general public, the controversy over the origin of life is not just an argument between conservative Christians and hard-nosed materialists.

> Some critics go so far as to liken Darwinism to creationism because of its slipperiness: it does not make specific predictions about what sorts of organisms evolution will produce, they charge, and so is never vulnerable to disproof. Like creationism, Darwinian evolution 'can equally well explain any evolutionary history,' says ichthyologist Don Rosen of the American Museum of Natural History in New York in a recent book. So heated is the debate that one Darwinian says there are times when he thinks about going into a field with more intellectual honesty, the used-car business. [43]

Hopeful Monsters for Wishful Thinkers

There's always the "hopeful monster" theory to dodge behind. Respected geneticist, Richard Goldschmidt, proposed it years ago. Supposedly two reptiles mated...an egg was laid...and when it hatched...out popped a parakeet! A similar origin is suggested for every kind of life on earth since no genuine transitional forms have ever been found. Recently a variation on this idea has become popular. They call it "punctuated equilibrium." Now that does sound more "scientific" doesn't it? All they're doing is theorizing ideas from a lack of evidence to support their primary belief.

The *L.A. Times* reported (6/25/78): "Scientists behave the way the rest of us do when our beliefs are in conflict with the evidence. We become irritated, we pretend the conflict does not exist, or we paper it over with meaningless phrases."

What is the real issue? The apostle Paul said it plainly in Romans 1:21:

They became vain [empty] in their imaginations [speculations]; their foolish heart was darkened; professing themselves to be wise [intellectual arrogance] they became fools [depraved and deluded].

When things don't add up, you either have to cover it up or change your mind. It's no wonder we're hearing some rather bizarre explanations nowadays. We should expect it.

The wisest man who ever lived (King Solomon) wrote:

"A fool's speech brings him to ruin. Since he begins with a foolish premise, his conclusion is sheer madness."
Ecclesiastes 10:12-13 (TLB)

THINK! Are the ridiculous escapes from reason just conclusions based on a foolish premise?

Let Us All Realize the Real Problem.

Scientist D.M.S. Watson wrote years ago:

"The theory of evolution is universally accepted not because it can be proved by logical, coherent evidence to be true, but because the only alternative, special creation, is clearly incredible!" [44]

"Clearly incredible?" Let's be bold enough to ask: "Which of these beliefs is really incredible?"

Life From Clay?

Now some scientists are saying that life evolved from clay! Isn't it amazing that some cannot separate the vast difference between the natural crystallizing properties of minerals and the infinite complexity of the living cell?

People once thought the earth was the center of the universe. Until the 1920s, respected doctors practiced "bloodletting" to treat sick patients. What has happened? Scientists have been proven wrong on many occasions. Often it takes generations to reverse strongly held theories.

Amazingly, leading **evolutionist** astronomer Robert Jastrow indicated that it will ultimately be necessary for science to come to terms with the supernatural. As he puts it: "For the scientist who has lived by his faith in the power of reason, the story ends like a bad dream. He has scaled the mountains of ignorance; he is about to conquer the highest peak; as he pulls himself over the final rock, he is greeted by a band of theologians who have been sitting there for centuries." [45]

"See to it that no one takes you captive through philosophy and empty deception, according to the tradition of men, according to the elementary principles of the world, rather than according to Christ."
Colossians 2:8 (NASB)

Evolution Opposes Christianity: There Is No Compromise

Evolution contradicts the biblical record of a finished creation.

The creative process is **not** now operating. God rested (Genesis 2:1-3). Evolution supposes a **continuous** creative process.

Evolution contradicts the doctrine of fixed and distinct kinds.

Evolution supposes a constant change back to a common ancestor. All flesh is **not** the same (1 Corinthians 15:38 & 39). Biogenesis (life begets life) is seen as "like begets like."

Evolution is inconsistent with God's omniscience.

It is all **chance** and **errors** in genetic translation to successive generations. God is **orderly** and is **not** the author of confusion (1 Corinthians 14:33).

Evolution contradicts the universal principle of decay.

Cursed is the ground (Genesis 3:17). The creation is waiting for release from death (Romans 8:21). The law of degeneration, disintegration, and increasing entropy is universally observed. The molecule-to-man philosophy is an illusion that ignores all the genuine evidence of nature.

Evolution produces anti-Christian results.

A corrupt tree cannot produce good fruit (Matthew 7:18). Evolution is at the very root of atheism, communism, relativism, racism, anarchism, and all manner of anti-Christian practices. Dangerous and deadly social problems are deeply rooted in the purposelessness of materialistic evolution: suicide, promiscuity, abortion, and chemical abuse, just to name the most obvious. Evolution has played a major part in reducing how humans think of themselves. Obvious problems are: low self-esteem, animalistic behavior, and depression due to a feeling of meaninglessness in life.

"THERE IS A WAY WHICH SEEMS RIGHT TO A MAN, BUT ITS END IS THE WAY OF DEATH." *Proverbs 14:12 (NASB)*

MIRACLE

Definition: "An event or action that apparently contradicts known scientific laws."

A Few Miracles Christians Believe In:

1. A Supernatural God exists ("supernatural" means existing outside the known laws of nature).

2. This Supernatural God has the ability to create the universe (all energy and matter) from nothing.

3. This Supernatural God has the ability to design an orderly universe.

4. This Supernatural God has the ability to create life.

Miracles Atheists Believe In:

1. Matter and energy created itself from nothing (violating the first law of thermodynamics).

2. Life originated from non-life (violating the law of biogenesis).

3. The universe began as disorder (the big bang) and became orderly over time (violating the second law of thermodynamics).

Both Christians and atheists have faith in miracles! What kind of "faith" does atheism require? What about Christianity?

Scientific Laws & Principles That Support Creation And Contradict The Theory of Evolution

Which Model of Origins Complies with the Natural Laws of Science?

Creation Model: What we observe today is the result of intelligent design, intelligent planning, and purpose. A Designer and Planner used means beyond the natural laws of science (supernatural). Plants and animals are offspring from parents of the same kind. They do not have a common ancestor.

Evolution Model: What we observe today is the result of chance events and long periods of time; there was no design, plan or purpose. Most (not all) evolutionists say there is no designer or planner behind everything. Everything originated through natural processes subject to the natural laws of science. The idea of supernatural intervention is rejected. Plants and animals descended from a common ancestor.

FIRST LAW OF THERMODYNAMICS:
This law states that in a closed system (such as our universe), energy and mass are conserved; energy and mass are neither created nor destroyed. **Energy and mass cannot originate from nothing.**

Evolutionists believe energy and mass did originate from nothing "billions of years ago." Evolutionism violates the first law of thermodynamics.

Creationists believe energy and mass could not have possibly originated from nothing by natural processes. Creationists say the origin of energy and mass was not natural but supernatural. See Genesis 1:34 & 14-18.

BIOGENESIS:
The fundamental principle of biology is biogenesis, which states that living organisms ONLY come from other living organisms (life comes from life).

Evolutionists say living organisms CAN and DID originate from nonliving matter by natural processes (remember, the first law of thermodynamics says mass and energy cannot self-create; now the evolutionist believes that nonliving matter and energy not only originated from nothing, but mixed together and started life!) "Life cannot arise by spontaneous generation from inanimate material today, so far as we know, **but conditions were very different** when Earth was only a billion years old" (quote from *Biology*, Nell Campbell, page 504, 1987). Evolutionism violates biogenesis!

Creationists say life cannot and never did originate by natural processes. Creationists say life originated from a supernatural source. See Genesis 1:11-27.

SECOND LAW OF THERMODYNAMICS:
states that with time, a closed system (like our universe–the earth is not a closed system which neither gains or loses energy) will become more random and disordered. Over time, energy will be "less available," (less able to do useful things like work). This law explains why we are running out of useful (not total) energy. For example, useful energy (gasoline) is converted to non-useful energy (heat).

Evolution Model: billions of years ago, the universe began from a few particles and evolved upward to make human beings. The universe became ordered and its useful energy increased by natural processes (which violates the second law).

Creation Model: the universe never became more ordered over time, and its available energy never increased by natural processes. The universe was created with order by supernatural plan (Genesis 1:1-31).

CAUSE AND EFFECT:
The most basic scientific principle, fundamental to all the branches of science and philosophy. Cause and effect requires that an observed event can be traced to an event that preceded it.

Creationists believe the universe has a "First Cause" (God). Atheists believe in NO "First Cause" for the universe. All scientists accept the principle of cause and effect EXCEPT those who reject a Creator.

Evolutionists think their belief is scientific and that the creation position is unscientific faith. In the study of origins, the evolution model clearly contradicts the natural laws of science, while the creation model complies with those natural laws.

Questions You Should Ask: 1. Why do you believe in the theory of evolution? (The BEST question to ask) 2. What is the best scientific evidence you have that supports the theory of evolution? Please don't be vague. (**Don't** be satisfied with vague answers like "because of biology and fossils." Try to get **specific** examples of why they believe it.) 3. Do you realize the evolution model for origins violates the laws of science? 4. What scientific laws do you know support the theory of evolution? 5. How did life originate? 6. Do you believe in spontaneous generation? 7. How did energy originate? 8. How did mass originate? 9. Is everything from design or accident? 10. What caused everything? 11. Is there a cause?

Scientific Law	Creation Model	Evolution Model
First Law of Thermodynamics	AGREES	CONTRADICTS
Biogenesis	AGREES	CONTRADICTS
Second Law of Thermodynamics	AGREES	CONTRADICTS
Cause and Effect	AGREES	CONTRADICTS

Exploring the Design and Perfection of the Creation

But now ask the beasts, and they will teach you; and the birds of the air, and they will tell you....Who among all these does not know that the hand of the Lord has done this, in whose hand is the life of every living thing, and the breath of all mankind?

Job 12:7, 9 & 10

The Truth about Origins Is an Appeal to Reason

When the ancient writer of the book of Job said, "Who doesn't know...," he was making it very clear that an open-minded discovery of nature should lead anyone to realize there is a Lord of creation responsible for the intricate design he sees all around him.

Let's examine a few of the amazing creatures and, as Job said, let them teach us about the awesome Designer who put them together.

The fact that the Creator made us conscious beings able to explore, to think, and to ultimately discover a personal relationship with the Creator himself, should cause every one of us to at least examine creation carefully for ourselves. We've got nothing to lose. When a scientist insists you can't discover what is absolutely true or that you can't consider God in a science class, he is writing his own rules.

Don't let such dogma mess with your mind.

Common sense can lead anyone to the truth if he is willing to find it.

That's why the Creator can be confident when He invites you to...

"Come let us reason together." *Isaiah 1:18*

Take a look at the picture of the pocket watch and a crude diagram of a living cell. Think about the complexity and purpose of each and then answer the following questions:

1. Which of the two is more complex?

☐ The watch ☐ The cell

2. Which needed an intelligent designer?

☐ The watch ☐ The cell ☐ Both

Remember! You're the scientist. First you observe. Then you suggest reasonable explanations. Then you test your theories. Then you make your conclusions.

With the advent of modern electricity, people marveled at ingenious advancements of things like the telephone, motion pictures, airplanes and refrigeration. But now we are humbled by the fact that the most high-powered mainframe computer is about as complex as tinker toys compared to the truly awesome eyes of a living octopus (which are very similar to human eyes).

No intelligent person would say modern technology happened all by itself. Yet how odd that otherwise sensible humans will insist that the wonders of the natural world originated without a designer because they can't conceive of Him. Why is this contradiction so common today?

Until about 200 years ago, most thinking people were convinced of the reality of the Creator-Designer just by looking at the world of nature. An Englishman by the name of William Paley wrote his famous book titled *Natural Theology* to advocate the argument that design requires a Master Designer. He wrote that if someone found a pocket watch lying on the ground, he would inescapably reach the conclusion it had been designed by a watchmaker. Paley reasoned that the order and design of the natural world also points to the existence of an omnipotent Creator-Designer who is Almighty God – the God of the Bible.

Charles Darwin began to popularize his theory of materialistic evolution by natural selection with his book in 1859. Paley's sensible argument became increasingly rejected by a culture wanting freedom from the moral restraints of a seemingly hollow state church in England. Philosophical debaters argued that design in nature was only apparent. What seemed to be wonderfully designed was simply the result of millions of years of chance processes. But did evolution really destroy the argument from design?

Don't Forget the Main Point Here...

What we're talking about in this whole design argument is **information**. If Darwin knew what modern science is only recently learning about the systematic coding of genetic information for every detail of biological machinery, the evolutionary argument might never have gotten off the ground. Molecular biologists like Michael Denton (whom we have quoted earlier) recognize that the world of the living cell, and its astounding miniature computer powerhouse that directs every detail, is extremely complex bio-machinery. The wonder of it all is the FACT that a living cell can do something that no man-made machine can do: it can make a perfect copy of itself in a few hours!

Human designers often patent their designs so others cannot steal their inventions. Yet most man-made machines and technological systems in our modern world have counterparts in the natural world. That means they are copies of a prior invention, and often poor copies at that.

Gecko's Toes Use Atomic Forces...

According to a report by scientists in the science journal *Nature*, geckos are small tropical reptiles whose ability to walk up walls and across ceilings has previously defied all attempts to explain how they could stick to any surface with no sign of any glue or suction mechanism to hold them on. A group of biologists and engineers studied the microscopic hairs on gecko toes and found the ends of the hairs directly attach to molecules in the walking surface by van der Waals force, a type of attraction between atoms. The report concludes that engineering a structure like the gecko foot is "beyond the limits of human technology," but scientists express the hope that the "natural technology of gecko foot-hairs can provide biological inspiration for future design of a remarkably effective adhesive."

If understanding how to manufacture gecko foot-hairs is beyond intelligent human engineering, it is certainly beyond blind, mindless chance. The more we look into the precision engineering found in living things, the more we are reminded of the apostle Paul's warning that human beings who study the created world are without excuse for ignoring the Creator.[46]

As Michael Denton points out: "The conclusion may have religious implications, but it does not depend on religious presuppositions."[47]

Nature's Chemists?

The Phenomenal Firefly

Have you ever seen a living light? In many places around the world people are fascinated by common little flying beetles often called fireflies or lightning bugs. There are many varieties and they don't look too outstanding, but wait until dark. What a show they put on. Their little taillights flash and glow brightly as they mysteriously communicate with many other fireflies nearby using their lights as signals. It is believed they use their lights in selecting mates and to attract others to a discovery of food. Of course other beetles do quite well without built-in light bulbs don't they? So why would a firefly evolve? Even more puzzling is the question of how a firefly could evolve!

The amazing ability of living creatures to produce light has been named bioluminescence. The firefly produces light by a complex process that scientists are only beginning to understand but cannot duplicate. A unique chemical named luciferin is combined with an enzyme named luciferase plus another compound produced by the bug's little body called ATP (adenosine triphosphate). Mixed in just the right proportions and in just the right timing, light is produced when oxygen is added to the formula.

The light from a lightning bug is sometimes called "cold light" because the energy conversion process is so efficient that no measurable heat is formed. Contrast this with our standard light bulbs that convert about 80% of the energy they use into heat. If it worked that way in a lightning bug, they would get quite toasty whenever they turned on the lights. And if you left the lights on too long… well, don't worry. The firefly was created perfectly able to avoid the problem of overheating.

How many everyday creations with complex chemistry do we take for granted?

Think of the chemistry that converts food into the muscles of your body, or that converts light entering your eyes into imagery in your brain! Complex chemistry is all around us as a testimony to God's awesome works.

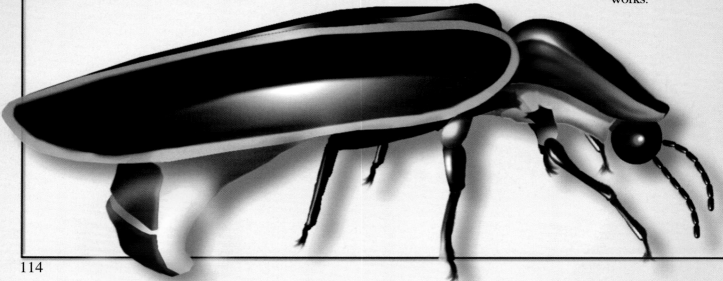

The Bewildering Bombardier Beetle

In the 1981 edition of World Book Encyclopedia's *Science Yearbook,* a fascinating article begins:

"The bombardier beetle is one of many insects that makes and uses complex chemicals to protect itself from its natural enemies."

You can identify this bug found under rocks near water especially by its beady black eyes. It has a yellowish-orange head, chest and legs, and greenish-blue wing covers over the back part. It's no more than three-quarters of an inch long.

Upon examining this creature in the laboratory, biologists have discovered that inside the bombardier beetle's body are two special chambers that produce two special chemicals, hydrogen peroxide and hydro-quinone. These chemicals are mixed and sent to a storage chamber that is connected to a second chamber called the explosion chamber, through a muscle that acts as a control valve. The explosion chamber has a number of small extodermal glands feeding into it that add an enzyme catalyst. When that happens, a violent cannon-like explosion is sent out the strategically positioned tube at the rear of the beetle's body and into the face of a would-be attacker. The hapless predator is left gagging in a steaming, noxious smoke as hot as boiling water, while the beetle scurries quickly for cover under the next rock. His special turret-like artillery fires with accuracy in whatever direction is necessary.

Witnessing a Real Miracle

Can you imagine (if evolution were true) the extremely delicate defense system of the bombardier beetle before it ever evolved to be fully functional? If the chemicals were not just the right strength or of just the perfect ingredients, can you see the picture of Mr. Bombardier Beetle manning his battle station for the first time? If it fizzled, his inadequate protection would be just so much extra baggage allowing his predators an advantage, and that would be bad news for Mr. B.B.

What about another possible hazard?

If the inner chambers or tubes weren't perfectly organized from the beginning, or if the control valve did not close when the explosion took place, think what would happen. He goes to take a deadly shot and it backfires! Mr. Bombardier Beetle is blown to pieces or has his innards cooked by his own defense mechanism and he never even makes it to the endangered species list.

You don't have to have a college degree to realize that a multitude of precision details had to be working perfectly from the beginning of this lowly bug's existence.

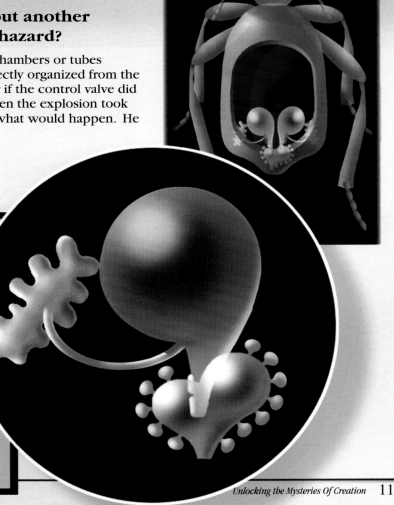

OK. You were right. We need to evolve an 'inhibitor'.

Evolutionary Adaptations Testing Grounds -- Top Secret --

Miracles on Wings

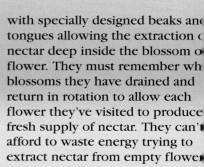

The Jewel of the Sky

If you ever watched the amazing gyrations of a hummingbird around a liquid feeder you can't help but sense a deeper meaning for the word AWESOME. With some 300 varieties worldwide, the hummingbird is the smallest bird in nature. Weighing less than a tenth of an ounce, he has much in common with a helicopter. These aerial acrobats fly backwards, forwards and sideways and hover in midair. Their wings beat an incredible 80 strokes per second and their hearts beat 1,000 times a minute. They inhale 250 times a minute and their metabolic rate is so high that it must feed almost constantly.

But, since there are no rods in the hummingbird's retina for night vision, its vital processes shut down to a state of hibernation at night. This is different from ordinary sleep, and unlike animals that hibernate in the winter only, the hummingbird hibernates every night – except when the female is nesting. Then, she stays awake to sit on her eggs. Every night the whole body slows down and the temperature drops to conserve energy.

The nest of the hummingbird is not much bigger than a postage stamp, made out of thistledown and cobwebs. And built into this pint-sized bird is one of the most complex flight mechanisms known. Their wings rotate at the shoulder socket, giving them their helicopter motion. To maintain this high energy output,

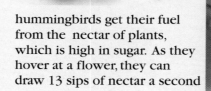

hummingbirds get their fuel from the nectar of plants, which is high in sugar. As they hover at a flower, they can draw 13 sips of nectar a second

with specially designed beaks and tongues allowing the extraction of nectar deep inside the blossom or flower. They must remember which blossoms they have drained and return in rotation to allow each flower they've visited to produce a fresh supply of nectar. They can't afford to waste energy trying to extract nectar from empty flowers.

What would you say if someone told you they believed a helicopter evolved without a designer? Knowing there is no other bird or pre-bird creature (living or fossilized) that even slightly resembles the amazing hummingbird with its integrated features, the sensible thing to do is simply admit that here again is another tell-tale example of the Creator's awesome design.

The Wonderful Woodpecker

The woodpecker is totally different from other birds. Every part of his body is especially suited for drilling into wood. His short legs and powerful claws are absolutely essential for holding on tightly to vertical tree trunks. They are very different from the spindly legs of other birds.

The beak of the woodpecker is also very unique. Banging away as much as a hundred times a minute, the woodpecker's beak has to be much harder than the beaks of other birds. Even more special is the way his beak is connected to his skull. There is a resilient shock-absorbing tissue between the beak and the skull that is not found in any other bird!

Or consider the woodpecker's tongue. This remarkable instrument is barbed in most of the two or three hundred species of woodpecker found on earth. It's about four times longer than the beak, wrapping around the back of the bird's skull in a very purposeful design. Some species of woodpecker produce a sticky substance coating the tongue for baiting ants. All of the woodpecker family uses this snake-like tool for penetrating deep inside a tree trunk to ferret out ants and grubs.

An Exacting Engineering Accident

After realizing just a few of the marvelous details of our little feathered "machines," try to imagine how such an invention could be engineered by a collection of rare and serendipitous chances.

Besides his powerful beak and special tongue, keep in mind that he also has a keen sense of smell. Together with his highly sensitive ears, Mr. Woodpecker can detect insects crawling around under the

bark of the tree he's scaling. By the way, how did the mysterious woodpecker learn to climb trees so well? His specially constructed tail feathers are stiff enough to actually brace him securely wherever he grabs a vertical perch. The engineering for such a technological wonder as the woodpecker boggles the mind.

Now, think about it. If the woodpeckers did evolve, can you imagine the obstacles the early ones would have had to overcome?

Think of the primeval headaches due to a lack of properly developed cushioning in the beak design! Worse yet, think of all the broken beaks that weren't hard enough to take all the jack hammering yet!

What about the primal woodpecker's legs and claws? In the early days, how on earth did the poor thing manage to hang on for dear life, risking everything for the high-class meals at the top? You can't help but wonder, "Why would this primitive daredevil bother to peck away for grubs in wood of all places?" After all, there have always been plenty of bugs crawling around on the dirt below.

It becomes quite apparent even to the casual bird-watcher that no supposed ancestral mutant would have ever survived to produce the marvel of God's creative genius we so commonly see today.

Lets see, I'll need to evolve a stronger beak, maybe some tougher toe nails... man I'm getting hungry!

It's What's Inside That Counts

What Wonders Show It Is Impossible for Birds To Evolve from Reptiles?

The most obvious difference between birds and every other creature is the intricate design of feathers. The engineering marvel of these structures is vastly more complex than the scales on a reptile. Do a little real science sometime by carefully observing a feather. See for yourself how feathers proclaim loudly that they were designed by a brilliant Engineer, and not by a materialistic process. And consider something else about birds that few people realize.

The breathing system of birds is uniquely different from the bellows action of simple mammal and reptile lungs. Though relatively small, bird lungs are connected to a system of inflatable air sacs spread through the chest cavity and even into the bones and breast muscles. Instead of inflating and deflating the lungs the way other creatures do, a steady flow of air efficiently delivers oxygen to the bird's blood supply by pumping the air constantly through the air sacs to the lungs. With each inhale and exhale action, the lungs remain full of air, so a bird is able to keep exerting tremendous amounts of energy without getting "winded" or left panting for breath. This is particularly important during high altitude and long migratory flying episodes, which of course, raises more wonders of design that have no naturalistic explanation… none.

It's just as the Bible says:

"… the invisible things of God from the creation of the world are clearly seen, being understood by the things that are made, even His eternal power and Godhead; so that they are without excuse."

Romans 1:20

An Inner Mystery of the Graceful Giraffe

When you see an 18-foot high crane-like giraffe, you can't help watching that amazing neck as it bobs up and down. The giraffe has one of the largest hearts in the animal world and nearly double the blood pressure of any other creature. When you consider what a stretch uphill it is to pump the blood to his head, you can see why there's a problem.

But what seems like a problem has a marvelous hidden solution. The jugular blood vessels in the neck have a series of one-way check valves to hold back the blood from rushing to the brain when it lowers its head, and to prevent it from flowing away from the brain too quickly when it lifts its head again. As an additional protection, a network of spongy tissue at the base of the brain soaks up any excess blood. Could you call this an amazing demonstration of plumbing technology? How important do you think this technology is to the giraffe?

Now, you know how it is when you have been bending down low and then suddenly stand and lift your head up. You get a little dizzy, don't you? Think of the giraffe. Can you imagine those valves not working perfectly at any supposed evolutionary stage? The first time he bent down for a drink of water the poor giraffe would literally blast his brains out from all that blood bearing down on his head. The simple act of raising his head up from a nap would result in such a loss of blood from the brain that he would pass out and be easy prey for a nearby lioness.

One of Darwin's silliest ideas in his book, *The Origin of Species*, was that during a long drought, some imaginary pre-giraffe that was taller than others could reach the scarce leaves upon which it fed. These survivors supposedly "left offspring inheriting the same bodily peculiarities, while individuals less favored in the same respects would have been the most likely to perish." And thus Darwin concluded, *"by this process long continued, an ordinary hoofed quadruped might be converted into a giraffe."*

You can't help wondering how the baby giraffes managed to survive during this incredibly long drought! Of course sensible scientists now know that acquired characteristics cannot be passed on to offspring. Furthermore, no fossil evidence supports the idea. Giraffes have always been giraffes, plumbing and all.

Again, the conclusion is obvious to one willing to discover the truth. As you continue to investigate creations as humble as the little honeybee or as grand as the marvels of our human bodies, the evidence of intricate design is everywhere.

O LORD how manifold are Thy works. In wisdom hast Thou made them all. The earth is full of Thy riches. **Psalm 104:24**

"The things that are made" tell us about the Creator who is the all-powerful designer of them all. That is why the Bible declares without apology, "they are without excuse." Think what that means. No one will ever be able to stand before God and tell Him there is a good excuse for not believing in and obeying the call of God. Every human who ever lived has had the clear evidence of the designs and intricacies of creation to loudly declare to him, "There is an Almighty God who made all this."

And as Jeremiah the prophet said,

"You will seek Me and find Me, when you search for Me with all your heart."

Jeremiah 29:13

Marvels from the Deep Blue Sea

Seahorse

For a fish, the seahorse has got to be one of the strangest. They don't look like a fish. They swim in a vertical position. They anchor themselves by wrapping the tips of their serpentine tails around seaweed. And strangest of all, the male seahorse gives "birth" to the babies.

That's right, every baby seahorse on earth "hatched" from a special pouch in its FATHER'S abdomen. The female seahorse deposits several thousand eggs at a time into the abdominal pouch of the male where they are fertilized. After that she swims off and has nothing more to do with her eggs or her mate. The "pregnant" male seahorse protects the eggs in his pouch, and when they hatch, a nourishing fluid is produced to feed the babies. Another two weeks passes before the male "gives birth" to a swarm of miniature seahorses that are immediately and completely on their own.

The leafy seahorse that lives near Australia is even more bizarre. Its body is covered with long waving bits of skin that resembles the look of the seaweed in which it hides. Its camouflage is almost perfect in hiding it from predators that would eat it if it could be seen.

How could male seahorses without a pouch, EVER evolve those WITH a pouch? Why would they bother to do it? How would the eggs survive before the females started depositing their eggs in the father's pouch? How did the nourishing fluid evolve? Who convinced the males they should stick to this new brooding job even if they had an unused pouch? Well, you can believe whatever you want, but you'll never find a convincing argument that can explain all this any other way than that it was all CREATED!

Sharks

People seem to be conditioned to fear sharks, but only a few of the 300 species of shark are any real danger to humans. They do look menacingly fearsome though, with their rows of pointed teeth, serrated like steak knives. We could wish we had the tooth replacement capability of sharks. When they loose or break a tooth, it is soon replaced by another fully developed tooth ready to flip from the rear of the old tooth and into place. In a shark's lifetime, it may use 24,000 teeth. No wonder shark teeth are so abundant in some of the fossil deposits of the world. One thing those fossils show us (again) is that sharks have always been 100 percent shark from the very beginning of their existence. There is no evidence to even suggest that sharks evolved from other creatures.

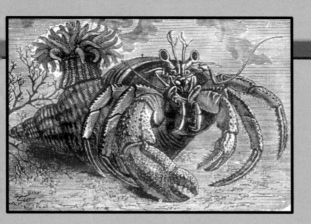

Hermit Crab

Of all the types of crabs on the ocean floor, only one has no protective shell of its own. The hermit crab carries around a hull borrowed from the ocean's scrap heap of old shells. This odd crustacean keeps replacing his old house as he grows, and uses his big pincher claw as a door to his little hideout.

Have you heard of "symbiotic relationships" between living things? The hermit crab has a symbiotic relationship with sea anemones. Anemones look more like a plant than an animal but they actually eat little creatures and they are quite colorful too. Their waving tentacles deliver a paralyzing sting at the slightest touch, but the hermit crab often is seen carrying one of these dangerous hitchhikers around, stuck on top of its temporary house. This curious cooperation works well for both creatures. The sea anemone eats food particles left by the crab, and the crab is protected from creatures that would like to eat it, like octopuses, who wouldn't dare disturb that numbing anemone.

There is zero evidence that the ancestors of hermit crabs ever had their own attached shells that they later lost. In fact, there

THINK! Why is the hermit crab a puzzling difficulty to an evolutionist?

would be no evolutionary advantage for a crab to lose its own natural shell and put itself in the vulnerable position of having to find a series of borrowed shells as it grows. To evolve a symbiotic relationship is no simple matter either. Evolution has no suggestion how this amazing programmed information could have arrived. "Survival of the fittest" has no explanation for the ARRIVAL of the fit. And hermit crabs have a perfect fit in their humble marine neighborhood. This is yet another effect that can only have been caused by a vastly intelligent and powerful Creator.

Sea Slug

The nudibranch sea slug looks kind of like a colorful blob with finger-like appendages growing on its back. This animal has a very peculiar dietary habit. You see, the sea slug's main food is sea anemones, which are covered with stinging cells. Normally those cells explode at the slightest touch and shoot a poisonous dart at any creature that happens to touch it. The sea slug tears sea anemones apart and swallows the stinging cells without exploding them. But even more amazing is that the stinging cells are not digested when they reach the sea slug's stomach. Tiny tubes, lined with moving hairs, link the stomach to the waving "spurs" on the back of the sea slug. The stinging cells swallowed by the sea slug are taken from the stomach, through the tubes, to the tips of the spurs. The sea slug then stores these stolen weapons to use in its own defense. It shoots them at any fish that dares to attack it.

This clever apparatus could not have been acquired gradually!

THINK! Could some magical "mother nature" have evolved this amazing system of stolen weapons?

Every component of this equipment had to be completely formed right from the beginning. The sea slug must have been designed and created by the Creator. No other explanation makes any sense.

Darwin wrote: "If it could be demonstrated that any complex organ existed which could not possibly have been formed by numerous, successive, slight modifications, my theory would absolutely break down." The sea slug demolishes Darwin's theory.

Koala

Who do you know with a "Teddy Bear?" The Australian koala, though not really a bear, looks like a live "teddy" and is one of the cutest creatures around. These gentle little guys aren't all so little. Some varieties are up to five feet long. The koala spends almost all its life in eucalyptus (gum) trees and is built with very special features. Long curved arms and sharp claws enable it to easily climb wide tree trunks. The vice-grip paws are a big help. Their first two toes oppose the last three, like the opossum's, but in a peculiar way. The first toe on the rear feet lacks the sharp claw of the other four toes, but clinches to them. It is used in a grasping thumb-like way on smaller branches. On the smooth gum tree trunks the widespread claws

of the koala dig into the bark like hooks.

The mother koala's pouch opens backwards toward the rear. Only wombats, bandicoots and Tasmanian tigers have pouches like that. Kangaroos and opossums have pouches that open front-wards or upwards. Now, a wombat is designed to avoid the pouch filling with dirt as it is burrowing. But why was the koala pouch built to enter from the rear? You might think this would risk babies falling out the back door when Mama is scurrying up trees, yet babies are always safe inside for the half a year they nurse milk inside the pouch.

Once weaned, the only food a koala eats is eucalyptus leaves,

but only a few species. They instinctively avoid the toxic ones. They get all their water from gum leaves and do not drink open water. Gum leaves are inedible and even toxic to most creatures, but the large intestines of koalas host microbes that help ferment the food, and their liver is able to detoxify the chemicals in gum leaves.

Nobody has figured how all these amazing features could ever have evolved by random mutations over a long time. And the funny thing is that perfectly designed koalas keep producing more perfectly designed koalas. None are born with evolving features to make it better adapted to its environment.

Pangolin

If you can imagine a cross between an anteater and an armadillo you get a little idea about the amazing pangolin from Africa and Asia. Here is a creature the size of a medium-sized dog covered in sharp-edged, bony plates of armor from head to tail. If threatened, the pangolin will roll into a tight ball so only its armor is exposed, making it nearly invincible.

After sleeping all day on a tree limb or in a deep burrow, the pangolin is equipped to find its food at night – termites and ants that are resting in their clay hills. With sharp beady eyes and keen noses they easily find plenty of ants and termites, helping to keep their populations in check. With powerful claws and legs, the pangolin breaks into a termite hill, and while the insects are racing around, the pangolin's long, sticky tongue laps up to a quart of them at a swipe. The stinging bites of ants don't bother the pangolin. He's equipped with transparent eyelids too tough to bite, and shut off valves at ears and nose to prevent ants from entering. His scales protect him well, but with no teeth he keeps pebbles in his stomach to grind up all his food.

The evolutionist would say that pangolins have "adapted" anatomical features and behaviors to fit its environment, but there is no evidence for any adaptive process that could create scales where there were none or new "valves" to close the door to its insect prey. Surely, any supposed "primitive" pangolin without the proper equipment would have never survived a night at the anthill "diner." To have all those curious features requires "programmed" information in its genetic code. This could only have been created by DESIGN, not by accident.

Platypus

You've likely heard of a duck-billed platypus. This furry little creature is one of the oddest in nature. It inhabits the waterways of New Guinea, Tasmania and southeastern Australia and is well known for having features of several kinds of animals. It has a bill like a duck, fur like a beaver, claws like a muskrat, short limbs like a lizard, cheek pouches like a monkey, and it lays eggs like a turtle.

The platypus actively searches for her food at night in the muddy bottoms of streams and ponds. It has small eyes and no external ears, but it has very good vision and hearing. Scientists eventually discovered something very special about the little platypus (up to 18 inches long when mature). They are equipped with electronic sensors in their leathery two-inch snout. While swimming underwater in search of their midnight dinner of shrimp, crayfish, and earthworms, their detector senses the faint electric waves produced by their prey.

The warm-blooded platypus is called a mammal but her mammary glands deliver milk to her babies in a rather peculiar way. The milk seeps through the skin onto the fur in special places where the babies discover they can lick her fur and be well fed. After some four months in the underground den with their mother, the offspring are ready to leave.

It doesn't take a genius to think about all these strange features and recognize the hallmarks of a very creative Designer. Yet for those who insist the platypus evolved from some kind of rat, there is no evidence that a platypus has ever been anything else but a platypus. How could a fleshy snout evolve into a leather bill? How could electric sensors magically appear where none existed before? How could it learn to swim so well, see so well, and eat under water also? And then of course, how does everything that evolving animals supposedly "learn" become part of their intrinsic INSTINCT? How do you program information that is learned into the genetic structure of offspring who inherit that programming from birth? Well, there's really only ONE ANSWER... it had to get there by DESIGN.

Unlocking the Mysteries of Original Man

"When I consider the heavens, the work of thy fingers, the moon and the stars, which thou hast ordained; what is man that thou art mindful of him? And the son of man that thou visitest him?"

Psalm 8:3-4

The Bible Model of Original Man

"And God said, Let us make man in our image, after our likeness: and let them have dominion over the fish of the sea, and over the fowl of the air, and over the cattle, and over all the earth, and over every creeping thing that creepeth upon the earth. So God created man in his own image, in the image of God created he him; male and female created he them."

Genesis 1:26-27

"And the Lord God formed man of the dust of the ground, and breathed into his nostrils the breath of life; and man became a living soul."

Genesis 2:7

According to the Bible, the fact of man's creation is clearly the result of a sovereign act of the almighty Creator. There can be no mistake that the wording of the Scripture and the understanding of conservative Bible scholars down through the ages agree that man's origin was supernatural and finished at his beginning.

Jesus further authenticated the truth of the Genesis account when He said, "Have ye not read, that He which made them at the beginning made them male and female" (Matthew 19:4)… and further in Mark 10:6 where He even narrows down the timeframe of that event saying, "But **from the beginning of the creation** God made them male and female." Notice He did not even slightly imply that the creation of man was set apart from the beginning of creation by a long period of billions of years.

I will praise thee; for I am fearfully and wonderfully made; marvelous are thy works; and that my soul knoweth right well.

Psalm 139:14

124

What Is the Popular Concept of Man's Origin?

According to the popular naturalistic belief about human origins, modern man gradually evolved from brutish "cave men." Those supposedly primitive ancestors had somehow evolved from some species of "pre-man" over an incomprehensible long period of millions of years.

A typical explanation of how this supposedly happened can be found in many colorful books. One book published by the Time-Life company wrote:

> Over ten million years ago a **versatile monkey** sired two distinct lines, the forest apes, and cave-camping pre-humans such as **Australopithecus** [literally meaning **"southern ape"**]. One of many branches of the Australopithicenes survived to become true men like Peking Man, a probable precursor of modern orientals.

So Is There Any Substance To The Idea of Human Evolution?

The search for fossil evidence to prove the theory of human evolution has gone on for over a century. It has proven to be a miserable failure!

In the first place, any bones of ancient "prehistoric" men are extremely rare. Yet, as we discovered in section one of our study, it would seem natural to find many burials of early man if humankind has inhabited earth for a million years or more.

Add to the lack of fossils the many recent admissions that evidence has been manipulated or ignored to generate public support for evolution. **Yet with almost total disregard for crucial facts, writers of popular books, magazines, and school texts strongly insist that man's evolution is a fact!**

THINK! As Paul the Apostle urged in 1 Thessalonians 5:21:

Prove all things!

Why Do People Believe That Humans Evolved from Animals?

The artists have colorfully illustrated the lineage of our supposed evolutionary family tree. Most people today are familiar with the monkey-to-man chart that decorates museum displays and evolutionistic textbooks. Let's go through the familiar chart, examine the names and the evidences used to support the theory, and see where it leads.

A colorful Readers' Digest book titled *The Last Two Million Years* features an introductory chapter that is quite typical of how the idea is presented to the public.[1]

> In Darwin's time there was little evidence to support his theory; but since then a whole chain of 'missing links' has been established by study of fossil bones found at prehistoric sites. The chart...shows how, over 40 million years, descendants of the early primates gradually evolved to produce modern man.

THINK! How can **"a whole chain"** be **"established"** when the links are **"missing?"**

THINK! How can any chart "**show how**" anything "gradually evolved" when the premise of evolution is and always will be only a **theory?**

And notice how 40 million years is rattled off as though it were unquestionable scientific fact! It appears that writers of such statements are, to say the least, overstating their case. But read on.

> ...the first breakthrough came when creatures adapted to standing and walking in an upright position...Most important, they were now able to make and use tools... distinguishing [them as] true man....

THINK! Knowing that **some animals** have the ability to **use tools** in some rather ingenious ways, why would we believe this is the trait "distinguishing true man?"

> As man's brain became bigger, responding to the demands of more complex hunting, he became taller, with more refined teeth and jaws.

Hold Everything! Did you get that? What, in the name of science, has the challenge of a complex hunting project got to do with the development of a larger brain? If that worked, you probably know somebody you'd love to send out on a hunting expedition, and maybe his brain would grow larger too! And if he became taller, what happened to all the short people on the planet? Furthermore, how does any of this produce teeth and jaws that are more refined?

Do the demands of more complex hunting really produce bigger brains?

Welcome to Never-Never Land… or maybe it's more like the Twilight Zone. But be careful. Words often don't mean what you think they mean. And just wait until you see how conclusions can be drawn from thin air when the evidence leads elsewhere.

Players in the Sideshow of Pseudoscience

When considering the subject of human evolution, it is very helpful to understand the human drama behind the conjectures that have come and gone over the years. Preconceptions and often a pathetically twisted antagonism to God or organized Christian religion have permeated the characters on this controversial stage. Unlike great men of science who have worked hard to unravel the mysteries of nature to benefit mankind or the environment, the protagonists of human evolution have had one goal in mind: to find evidence that rejects the God of the Bible and the Genesis account of man's creation. Their negative aspirations have fostered a world of speculative pseudoscience used to support various philosophies of meaninglessness and, worse yet, social chaos. This ungodly approach to anthropology is not an exercise in objective science, but an egocentric competition to reduce man to the level of a beast.

Recommended Reading: *In the Minds of Men* by Ian Taylor. It is a classic thorough study behind the scenes of the people, the events, and the findings of this modern mind-game that has confused the thinking of millions of people.

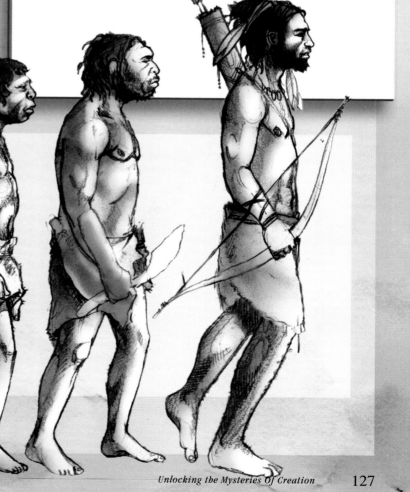

Where Are the Missing Links?

The Evolutionary Adam?

In the chart found in the book, *The Last Two Million Years,* we find the first entry labeled simply, "Common Ancestor."

The caption declares:

This creature is believed to have been a forest-dwelling creature, the ancestor from which modern apes and man both descend. **No traces of such a creature have yet been found.**

You're probably wondering why the publisher paid the artist to paint a picture of a supposed ancestor that has never been found. Good question. This is a good reminder of the fact that this religion of evolutionary origins is based on beliefs and pre-suppositions. So if there are no facts they take the liberty to put in a good drawing.

Other charts have started with a creature named "Gigantopithecus" (meaning "giant ape"). However, when enough scientists insisted it was nothing more than just that, a giant ape, it was finally removed from the line leading to man. Richard Leakey, the famed anthropologist, showed in his later books that Gigantopithecus simply became extinct and did not evolve toward man. (Note his article in *Time* magazine 11/7/1977).

If the eminent evolutionary expert on human origins has decided to take our "common ancestor" off the chart, we might as well remove it too. Since this "link" has never been found, we can only conclude that **it must still be missing!**

Ramapithecus

The chart in our impressive Reader's Digest book describes the next so-called "link" as "a more advanced primate... appearing by 14 million years ago." The interesting thing about *Ramapithecus* is that **it was all made up from one tiny piece of a jawbone about two inches long!** The "find" was made in India in the 1930s. Some time later another small piece of jawbone was dug up in Africa. They claimed it belonged to the same species.

Even though the evidence was fragmentary, some people can make a little of anything go a long way! Look at some of the evolution books on the library shelves from the 1950s to 1970s and you'll see drawings of *Ramapithecus* with all the details. His posture, length of arms and legs, shape of head and amount of body hair are all artistically complete. And just think: all of this from a couple inches of jawbone!

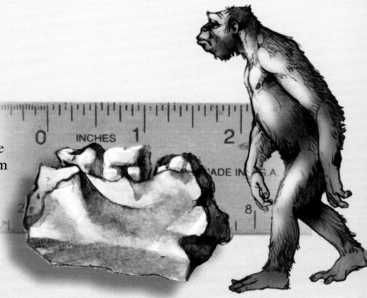

So why include this flimsy evidence in the chart? The answer is simple. Yale University paleoanthropologist, David Pilbeam, **believed** this was more man-like than ape-like! Now they realize that some living baboons have simlar tooth and jaw structures. So, it isn't surprising that *Science Digest* published this statement in 1981:

> A reinterpretation of … [this] … jaw … now suggests that *Ramapithecus* was an ancestor of neither modern humans or modern apes. Instead Pilbeam [himself] thinks it represents a third lineage that has **no living descendents.**[2]

If an old bone has "no living descendents," that means it can't be related to man. So now we have to remove another "link" from the chart. It's definitely a missing one.

Australopithecus

One of the long-standing and well-known suspects on the familiar line-up is called *Australopithecus*. The name literally means "southern ape." That should tell you a lot!

Quite a few skulls have been found over the years and given this designation. Perhaps the most famous one was found by Dr. Louis Leakey in 1959 in the Olduvai Gorge of Tanzania. He called it *Zinjanthropus bosei*, but it was dubbed "nutcracker man" and claimed to be 1.8 million years old. Two types of this extinct ape are included: *A. africanus* and *A. robustus*. Since the brain case and skull form of this animal are distinctly ape, you may wonder why it was included in the chart of man's ancestry.

The reason *Australopithecus* was included on the lineup evolving to modern man is simple: tiny **supposed stone tools were found near the bones.** When you see the pictures in the old evolution books, you'll notice the so-called "tools" are chips of rock about the size of a small acorn. What kind of micro-technology is supposed to be indicated by these pathetic bits of stone? Anybody would be justifiably skeptical about such conclusions.

It's interesting that Richard Leakey (Louis' son), in his book called *Origins*, has removed *Australopithecus* from the chart leading to *Homo sapiens*. The southern ape has been placed in a totally separate lineage altogether. If Leakey, one of the leading evolutionists of the 20th century, has removed this supposed "link" from the chart, we may as well remove it too, don't you think?

Lucy

The December 1976 issue of *National Geographic* magazine featured what they thought was a major discovery in Ethiopia. It was a collection of bone fragments from a three and a half foot tall chimpanzee-like skeleton found in 1974 by a young American graduate of the University of California at Berkeley, Donald Johanson. The bones were claimed to be over three million years old. They named it Lucy because the team was hearing their radio playing a rock and roll song by the Beatles titled "Lucy in the Sky with Diamonds." From about 1979 it gained the favor of many evolutionists as the key ape-like ancestor of modern man even though it is acknowledged by many to be a chimpanzee, but one that is claimed to have walked upright.

Now there are upright walking chimps living today, but how did they know this old chimp Lucy walked with an erect posture? The evidence was dependent on an interpretation of a knee joint. After a university lecture in Kansas, a well-informed creationist, Mr. Tom Willis, asked Mr. Johanson publicly where he found that important knee fragment. The answer: a mile and a half away from the rest of the skeletal fragments in a rock strata 200 feet deeper! Next question: Why include a fossil fragment so widely separated from the main find? Johanson insisted, "anatomical similarity" was all the justification needed.

THINK ! Is that kind of evidence adequate to make Lucy our ancestor? Such evidence sounds embarrassing. No wonder the popular evolution-sympathizing press has not made it public. It's likely just a matter of time before Lucy is also knocked from the branches of man's family tree.

This evolutionist display shows Lucy's chimp-like bones in a misleading human outline? More baloney!

The Discovery That Rattled
All the Other Bones

Homo Habilis (1470 Man)

Homo habilis is next on our chart from the Reader's Digest book. In June 1973, the *National Geographic* magazine published an article that was devastating to conventional ideas about human evolution.

It reported a new find in Kenya, Africa by anthropologist Richard Leakey, the leading evolutionary expert on the so-called "hominid" ancestors of *Homo sapiens*. The discovery was called "skull 1470" (fourteen-seventy) for its catalog number in the Kenya national museum.

Leakey made an astounding challenge, highlighted prominently in bold letters by *National Geographic*. He wrote:

"Either we toss out this skull or we toss out our theories of early man."

The anthropologist said this fossil was 2.8 million years old, yet it belongs to man's genus. In other words, Leakey claimed it was more man-like than any of the other near-man relics on the chart. The problem was that the skull was found beneath volcanic ash that had been acceptably dated for years by evolutionist reckoning as 2.6 million years old. That would make a human looking ancestor over a million years older than our nearest ape-like ancestor.

It's no wonder Leakey made the puzzling statement: "*It simply fits no previous models of human beginnings.*" And because of the skull's "*surprisingly large braincase,*" Leakey shockingly admitted, "*it leaves in ruins the notion that all early fossils can be arranged in an orderly sequence of evolutionary change.*"

Keep in mind that the National Geographic Society is a major financial supporter of field explorations (including Leakey's) to find fresh new specimens to put on the line-up of man's evolutionary origins. It is worth noting that they are willing to publish such discoveries with considerable fanfare even when they are controversial.

Let's realize the implications of Leakey's comments. He stated that the chart with which we have all become familiar is now a "*notion*" left "*in ruins.*" The "*orderly sequence of evolutionary change*" apparently does not rate any better than a "*notion.*"

After this high profile publicity, Leakey lectured in San Diego, California as well as other places. Audiences heard him explain his conviction that his discovery eliminates everything we've been taught about human origins. He said he had nothing to offer in place of the popular concepts.

What about the Artist's Reconstruction?

Though skull 1470 has a cranial capacity well within the range of modern humans, isn't it remarkable how the *National Geographic* artist can characterize a face he has never seen? All the soft superficial fleshy parts of a face are a big guess. Yet the ears, lips, nose, hair and skin color are all presented to the gullible public as though the scientists had a crystal ball into the past.

What Does Science Really Tell Us?

In his thoroughly researched book, *Bones of Contention,* author Marvin Lubenow brings to light that the facial bones were not clearly connected enough to know for sure if the face should be flat like a human or with jaw extended like an ape.[3] As he further pointed out, "*Homo habilis* is a flawed taxon, or category, because it is a mixture of fossils that can legitimately be called human, and other fossils that are definitely not human." Well now we have a problem. Evolutionists can't have a candidate for a missing link that is admitted to have a skull qualifying as modern man, but that dates back to over two and a

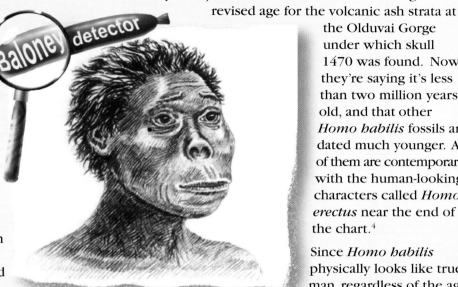

Artistic reconstruction of "1470 man"

half million years ago. This paradox continued for almost a decade.

Finally, in 1981, evolutionists came up with a technical way to adjust the radiometric date and assign a revised age for the volcanic ash strata at the Olduvai Gorge under which skull 1470 was found. Now they're saying it's less than two million years old, and that other *Homo habilis* fossils are dated much younger. All of them are contemporary with the human-looking characters called *Homo erectus* near the end of the chart.[4]

Since *Homo habilis* physically looks like true man, regardless of the age they assign it, how can it be something evolving to man? What would they do if they found human looking bones in the same geologic age assigned to the dinosaurs? Will they push man's origin back 100 million years or figure a way to reassign the age of the rocks again? Just wait and you will learn about even more mysterious finds (mysterious only because they don't fit the commonly accepted evolutionary beliefs about human beginnings).

So let's recap our chart of missing links. The three figures on the left have been removed for various reasons. The fourth is considered practically modern. Where does that leave us?

Is Homo Erectus the Missing Link?

The next "missing link" on our classic lineup is called *Homo erectus*.... Why? The Life Nature Library volume titled *Early Man* begins chapter 4 with the title: "*Homo erectus*: A True Man At Last!" The "erectus" part of his name means he walks with upright posture, unlike the apes that use their arms in walking. Two discoveries are often shown as examples:

Java Man

In 1887, a young Dutch medical doctor, Eugene Dubois, joined the army so he could be assigned to the Dutch East Indies (now Indonesia). His goal was to find the "ape-man" concocted by his professor, Mr. Haeckel (remember him?). By late 1891, his army crew of up to 50 laborers had extracted tons of animal bones on Java Island. Finally they brought him a tooth and the dome of a skull they found by the bank of the Solo River. Dubois declared it had both human and ape features. Almost a year later a human leg bone and another tooth were dug up 50 feet from the skull cap.

Dubois consulted with Haeckel, put the head bone together with the leg bone, and called it *Pithecanthropus erectus* ("upright ape man"), but it soon became known as "Java Man." With no qualifications to peg the age of the strata, he claimed his "missing link" was half a million years old. A 1907 German expedition discovered that the bone-bearing sediment was produced by a volcanic eruption. The locals reported that flooding had changed the river course in the 13ᵗʰ or 14ᵗʰ century. Thus, the bones could be as recent as 500 years. [6]

Java Man provoked controversy. Respected scientists of the time had trouble with Dubois' ego and eccentric claims. Dubois boasted his discovery of the ape man, and when contemporary experts disagreed, he finally refused to let anyone inspect his fossils. In 1920, he finally announced his discovery of two human skulls, known as the Wadjak skulls that had been found over 30 years earlier in Java and dated at 10,000 years old. Why didn't he expose those when he found them? Because they were human! And remember, his goal was to find the "ape man."

Since 1950, evolutionists have treated Java Man as the primary example of *Homo erectus*, an early form of man. But in Dubois' final years, trying to separate his find from other discoveries vying for fame, he insisted his "*Pithecanthropus*" was a giant gibbon, although he still called it the "real missing link." [7] It's another case of a "missing link" all right. There's no evidence here to link man to apes at all, but it still finds an honored place in many textbooks.

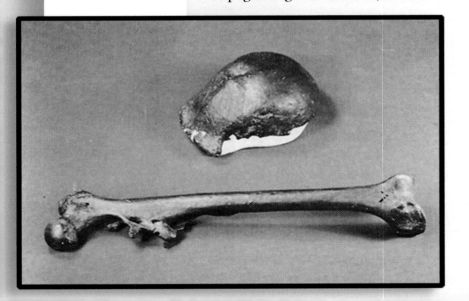

Peking Man

It's amazing what people see when they put great effort in trying to find something that supports what they are eager to believe. In 1921, 25 miles from Peking, China, two molar teeth were dug from a limestone hill named Chou K'ou Tien, meaning "dragon-bone hill." In 1927, another tooth was found by Davidson Black, a Canadian doctor. Excited to find the missing link in China, he announced his discovery as *Sinanthropus pekinensis*, or "Peking Man."

With a sizeable Rockefeller Foundation grant, scores of laborers were hired to sift through hundreds of tons of earth. In 1929, an almost complete brain case was found similar to Dubois' Java skullcap. By 1934, thousands of animal bones were found, including elephants and deer. **Mixed with them** were only 14 skull fragments of Peking Man along with 11 jawbones, 7 thigh fragments, 2 arm bones, a wrist bone and 147 teeth. When Dr. Black died, Franz Weidenreich took charge and fashioned a model skull that became known as "Nellie" from all the fragments. Casts of this composite reconstruction appear in museums worldwide. But what's the rest of the story?

In an effort to preserve and move the bone fragments during the confusion of World War II, **all of them were lost except for two teeth.** Before the war, the eccentric Roman Catholic priest, Teilhard de Chardin, invited French scientist Marcellin Boule, to visit Dragon-bone Hill. Boule published his opinion that the bones were obviously a collection of battered monkey skulls. After examining the bones and the entire site, he concluded that all the animal fragments were discarded by hunters who had eaten them.

Earlier still, in 1931, another French expert on the Old Stone Age, Henri Breuil, spent 19 days at the site, also at the invitation of de Chardin. Breuil found a 23-foot deep ash heap surrounded by fused soil, indicating a significant furnace used by an industrious people. Yet the evolution textbooks mention only "traces of fire" with the assertion that Peking Man was one of the most primitive users of fire.[8]

The 160-foot deep quarry is filled with broken animal bones from top to bottom. Also found there are the bones of normal humans, as well as man-made stone tools. These facts are rarely revealed in popular evolution-biased textbooks today. The details of the people and events behind all of this read like a murder mystery. Anyone wanting the full story will enjoy the account researched by Ian Taylor in his excellent book, *In the Minds of Men.*

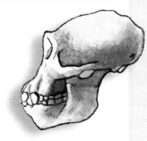

Ape

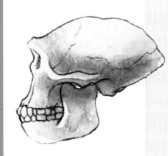

Homo Erectus

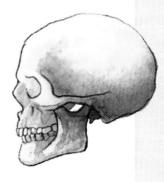

Modern Man

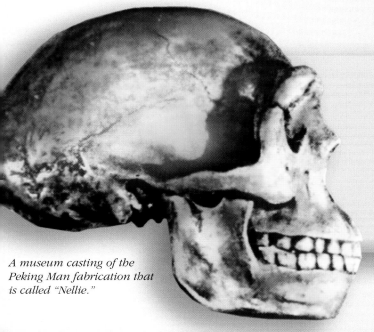

A museum casting of the Peking Man fabrication that is called "Nellie."

THINK ! What kind of science depends on models of evidence that are now lost?

After the death of all the main characters involved in Peking Man, he was re-designated as *Homo erectus*, just like Java Man. Yet the more than 200 other examples of *Homo erectus* found in Asia, Africa and Australia seem to physically qualify as thoroughly human.

Are Cavemen Our Ancestors?

Before we get to modern man on the far right side of the chart, we encounter two other individuals that are curiously labeled "men." This is appropriate since both Neanderthal Man and Cro-Magnon Man are both true humans indeed.

THINK! If two of our supposedly evolutionary ancestors are already evolved into man, then why are they in a chart of so-called "missing links" that is supposed to show us the creatures leading up to *Homo sapiens*?

Neanderthal Man

The very name, Neanderthal, arouses involuntary thoughts of a hunch-backed, primitive brute with a thick overhanging forehead and a gorilla-like face. But what is the real story on Neanderthal Man? (The proper German pronunciation is nee-ANDER-tall).

The Beginning of Fossil Man Deceptions

Remember that in the mid-19th century modern world, the prevalent understanding of man's ancestry was based on the Genesis history of creation and Noah's flood. Darwin's outrageous book was published in 1859. Promoters of Darwinism were under pressure to find evidence linking man with animals. Although the first Neanderthal bones were called archaic humans, their heavy features and bowed leg bones soon sparked the imagination of those wanting a missing link.

Following the first Neanderthal discovery, over 60 more similar fragmentary skeletons were found in different parts of Europe, Asia and Africa. Even female and child skeletons displayed evidence of a very strong breed of people. By the early 20th century, popular books commonly showed Neanderthal Men as naked, hairy, club-swinging, dim-witted brutes. But were they?

How did we figure Neanderthals were primitive?

From the 1920s major museums featured full-scale dioramas depicting Neanderthal families as primitive, backward and powerful. The public image of earliest man digressed. Instead of thinking about Adam falling from perfection in Paradise, modern students were impressed that we arose from brute beasts. People began joking that young men with physical prowess were akin to strong but stupid apes.

Professor Rudolf Virchow of the University of Berlin (who came to be known as the founder of modern pathology) studied the bones of Neanderthal Man in 1872. He saw evidence that here was an essentially modern *Homo sapiens* who suffered rickets in childhood, arthritis in old age, and had been struck more than once in the head before his death. However, another professor of anatomy interpreted the bones with an evolutionary slant and gave them the name *Homo neanderthalensis*.[9] However the evolutionary myth about Neanderthal flourished for a hundred years.

Finally, in 1981, articles began appearing in the public media that admitted what the academic elite had known for years. One headline read: ***"Neanderthal Man: He may not have been the hairy ape we thought he was."***[10] So why were Neanderthals displayed as hunchbacked and retarded looking? As the article said, "**one skeletal find turns out to have been severely deformed by age and arthritis.**"

If you were to give Mr. Neanderthal a shave, put him in a business suit, and send him downtown to pay the bills, he wouldn't stand out from the crowd at all. In fact, you've likely seen individuals on the street that looked a whole lot more primitive than Mr. N.

A really surprising fact about Mr. N that few ever hear is that his brain size was 10% to 15% **larger than the average modern human.** That sounds like evolution in reverse. [Though brain size used to be an indicator of evolutionary advancement, science now realizes that it has little to do with intelligence].

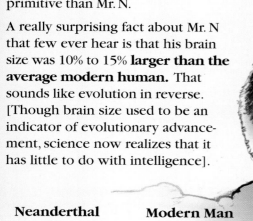

Neanderthal   **Modern Man**

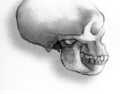

It's clear now that these people of ancient Europe were truly human in every way, and superior to us in strength. Even some of their social complexity can be learned from their burials, and there is no reason to depict them running around naked like animals. So is he an evolutionary link?

Neanderthals were once thought to have been our ancestors. Modern evolutionists tend to think he was a sideline cousin of man's ancestry that became extinct because of environmental factors or else was destroyed by whatever ancestor finally did rise to become man. But one thing is certain; Neanderthal is not a proof that humans evolved. This missing link is also **still missing.**

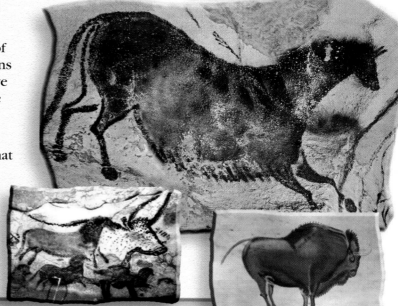

The Surprise Discovery of Cro-Magnon Man

In 1940, some boys were running with their dog in the countryside near Lascaux, France. The dog fell into a crack in the ground. When the boys rescued their pet, they prodded their way into an ancient cavern several hundred feet long. The walls were covered with colorful paintings of horses, deer, and bison.

These paintings are now famous – the skillful artwork of people we call Cro-Magnon (KRO-man-YO), which means "great big." Skeletons were found buried in another cave at Les Eyzies, France in 1868. Over 70 French sites have been found with Cro-Magnon art.

Evolutionists suppose these people go back 12,000 to 30,000 years, but they have no absolute way to verify that age. Making fancy paintings in caves doesn't make them less human. Do some people live in caves today? Is their artwork as fine? Actually, these Cro-Magnon artists were more skilled than most humans living in caves today. No wonder *Smithsonian* magazine titled an article: "**Cro-Magnon hunters were really US, working out strategies for survival.**"[11]

The Final End of All Supposed "Missing Links"

Now that we've dug out the facts behind the so-called missing links, it looks like our classic evolutionary chart needs some major revisions, doesn't it? There's modern man walking off the right side and all the others are just phantoms. Remember the book that declared these characters were "**established** missing links"? Truth is…they're still missing! Why? Because they were disqualified as "apes", "men" or "???." There have been other famous "links" on the chart in the past, but they've also been removed. Let's examine a couple…

Piltdown Man

The director of the Natural History Museum of London, Arthur Woodward, declared a historic discovery in 1912. A Doctor Charles Dawson unearthed a human-looking skull cap and an ape-looking jawbone from a gravel pit near Piltdown, England. It was called "Dawn Man" and proclaimed to be 500,000 years old! The scientific community was convinced. Now they had proof of a transitional creature between ape and man.

Over the next 41 years, some 500 academic dissertations were written on the famous Piltdown Man. Then in 1953, some scientists finally performed modern chemical analysis on the fossils. They confirmed serious critical reports made years earlier. They found the teeth had been filed to fit, and the bones had been stained to make them look old. The whole thing was a fraud! A Fake! They wanted to believe it so badly they taught it as fact to a whole generation of school children. By the time this hoax was exposed, the public had gotten used to the idea that science had really proven the ape-man story.

The Piltdown Hoax

"The reconstructed skull (left) shows bone in brown and filled-in clay in white. The sculptured face on the right appeared in museums all over the world for decades before it was finally admitted to be a fake."

THINK! What valuable lesson can we learn here? The whole episode proves that scientists aren't infallible. Their preconceptions can lead them to be deceived, so they are prone to believe untrustworthy data… and like a lot of people, they tend to find what they want to find, whether it is genuine or not.

Nebraska Man

The paleontologist who headed America's Museum of Natural History in the 1920s was Professor Henry Fairfield Osborn. He was a confirmed evolutionist, bent on battling the famous Christian defender, William Jennings Bryan, over the teaching of evolution in public schools. In 1922, Osborn announced the discovery of a fossil tooth from Nebraska. He claimed it was an early ape-man, whom he named *Hesperopithecus haroldcookii* in honor of its discoverer. It soon was popularly known as "Nebraska Man."

Dental experts at the American Museum of Natural History studied the fossil tooth carefully and concluded it was from a species closer to man than ape. These evolutionistic experts delighted in the first North American discovery of a missing link, especially because it was found in Bryan's own home state – Nebraska.[12]

An Englishman who was involved in the Piltdown discovery a few years before, persuaded the widely read *Illustrated London News* to publish an artist's rendering of Nebraska Man and his mate. A full two-page spread was drawn and distributed worldwide, showing a naked pair of stupid looking ape-people. The club swinging male no doubt impressed a whole generation with the idea that human ancestors were far inferior to the biblical Adam and Eve.

The year 1925 is memorable for being the year of the world's most publicized court trial in history – The famous Scopes "Monkey Trial." The American Civil Liberties Union pressed for the trial so they could make a public spectacle of the evolution controversy. In those days, creation was taught in American public schools, and teaching evolution as scientific fact was illegal in the state of Tennessee. Though the ACLU lost the case, they eventually won the battle since evolution came to be the only origins explanation accepted in public schools by the 1960s.

During the Scopes trial, the impressive picture of Nebraska Man was well planted in the public mindset. Along with Piltdown Man and the warped views of Neanderthal Man being popularized at that time, young students were being led to believe that science had evidence for the evolution of man. You can see how this pushed the biblical view of Adam and Eve into the realm of mythology.

In 1928, just three years after the Scopes trial, scientists discovered a mistake. It wasn't a tooth from an early human at all. **It was a tooth from an extinct pig!** In 1972, living herds of the same species were discovered in Paraguay. Yet the *Encyclopaedia Britannica* (1929, vol.14, pg.767) explained that *Hesperopithecus* was "a being of another order."[13]

Mark Twain wrote a rather revealing insight about such cunning professional deception: "There is something fascinating about science. One gets such wholesale returns of conjectures out of such a trifling investment of facts."[14]

Think! Was the public really made aware of the extreme misrepresentation that had been made about the ancestors of humans?

How Long Has Man Been Truly Human?

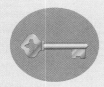

Is there **any** evidence to link man to ape-like creatures? The cover of *Time* magazine (November 7, 1977) headlined "How Man Became Man" and featured anthropologist Richard Leakey beside a live made-up model representation of *Homo habilis* or "1470-Man." The naked black model is wearing an imaginative mask supposedly showing us the look of this questionable "link."

Candidly, the author of the article for *Time* makes this stunning confession:

> Still doubts about the sequence about man's emergence remain. Scientists concede that **their most cherished theories are based on embarrassingly few fossil fragments** and that **huge gaps exist in the fossil record.**

Though many fossil bones of true men and true apes have been added to the catacombs of museums around the world since Neanderthal's discovery, it has been observed that if you put all the really meaningful so-called **hominid relics** together, **you wouldn't even fill a single coffin.**

The whole basis on which paleontologists classify fossil apes and humans is misleading. The time has come to admit that the system by which we name things is inadequate in dealing with things that have a time dimension.

The finds to which Leakey refers come from East Africa, as do so many of the discoveries of recent years. "Footprints In The Ashes of Time" was the title of the article featuring these tracks in the April 1979 issue of *National Geographic* magazine.

The volcanic ash in which the prints are found has been dated by the potassium argon method. Keep in mind the implications of that as described earlier. The tracks of many animals were also found there, "frozen" as it were, in cement-like mud during a volcanic disaster.

So Where Is The Oldest True Man?

An interview with Richard Leakey was printed in the *Vancouver Sun* (March 19, 1982). According to Leakey, man's ancestors go back 3.75 million years to fossil footprints discovered by his mother, Mary Leakey, in Laetoli, Kenya.

This stunning conclusion relies of course on the doubtful ages determined by radiometric methods. But the surprise is that this discovery pushes man's earliest supposed man-like ancestors back before more primitive creatures which were once thought to be our forefathers.

The article quotes Leakey as saying:

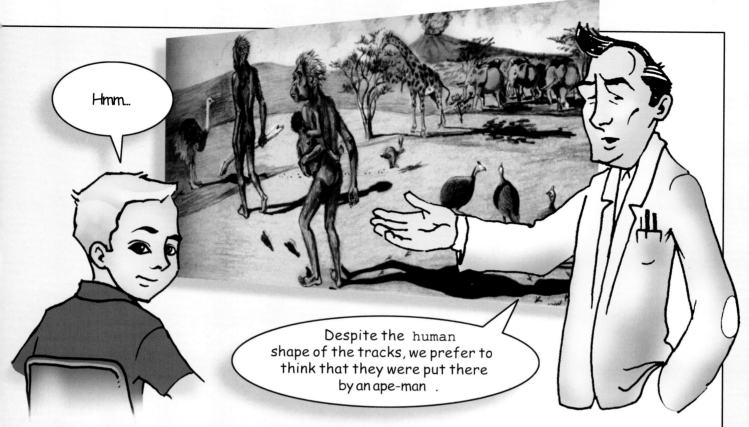

Hmm...

Despite the human shape of the tracks, we prefer to think that they were put there by an ape-man .

The human prints are exactly like ones you might make the next time you walk bare-foot along a lakeshore. What do they tell us?

Expert trackers identified that the animal tracks were associated with a variety of modern animals. A talented artist was commissioned to recreate the scene for the magazine. It's interesting to notice the modern appearance given by the artist to these animals. The guinea fowl in the painting are like those living today. The giraffes are also modern. The elephants look just like the ones in a modern zoo. So do the ostrich and the hare. But when you come to the human tracks in the painting how has the artist portrayed the individuals who are making them?

The feet look reasonably man-like, but as you look higher up the figures you get the distinct impression that the artist is doing a bit of imaginative embellishing. Does he know something we don't? In these distinctly human tracks the artist has placed an ape-man – some kind of half-and-half creature that no one has ever seen. Now you would think they could figure out what kind of creature makes human footprints. But when you see the volcano erupting in the distance and the ape-men sauntering along and looking the other way, you wonder if the artist really has a track on reality at all.

Though no one has ever found any creature except a human, able to make a human footprint, such a fact is irrelevant to this article. It is very clear that artists have done as much to formulate the public view of

man's evolution as any scientist ever could.

A New Theory of Evolution

Newsweek magazine carried an article on March 29, 1982, with a new twist on human evolutionary theory. Since the fossil finds are all either apes or men, something has to be done to explain the mystery. So now they have a new theory.

> Instead of changing gradually as one generation shades into the next, evolution as [one Harvard scientist named Stephen J.] Gould sees it, proceeds in discrete leaps. According to the theory of punctuated equilibrium there are no transitional forms between species, and thus no missing links!

How convenient!

It was once said that evolution happened so slowly that no natural examples could be found to prove it. **Now some say that it happened in such quick "leaps" that no fossil evidence (links) could be found to prove it.**

The apostle Paul, in his letter to the Roman Christians writes:

> **They knew God but they didn't thank Him… Their thinking became futile and their foolish heart was darkened… Though claiming to be wise they became fools… and exchanged the truth of God for a lie.**

Where does the real evidence lead?

What Makes Man Truly Human?

There is **no evidence in the real world to even suggest,** let alone prove, that man has ever evolved from some lower kind of animal!

All fossils that have ever been found (in this arena of study) **are either all ape or all man or fake!** No real fossil has been proven to be transitional. In other words, **there are no missing links!**

The truly wise man will say with the Psalmist, David:

"I will give thanks to Thee, for I am fearfully and wonderfully made; Wonderful are Thy works, and my soul knows it very well."

Psalm 139:14

Contrary to evolutionary thinking, man was made completely human from the beginning. He was a perfect specimen of human complexity. He needed nothing more to be fully functional.

What makes man different from all the other creatures God made?

- It is not his anatomy, though that is unique!

- It is not his ability to walk upright, though that also is unique!

- It is not his ability to make and use tools!

- It is not really even his verbal speech, though that too is also uniquely man's gift!

- It is not just his great intelligence, as wonderfully superior as that is!

Man is unique because of his deeply spiritual nature....

Made in God's Image

In Genesis 1:26 God said, *"Let us make man in our image according to our likeness...."*

What is God's likeness? God is not a man of flesh and bone. As Jesus said in John 4:24, *"God is Spirit...."* And that is primarily the nature of man that makes him distinct from all other creations of God.

Unlike all the animals, God made man in the class of spirit beings. Yes, man lives in a body of flesh, but his eternal, God-made spirit makes him the highest and only creation who can operate beyond the limits of mere fleshly animals.

How close to God is Man?

In Psalm 8:5, we read that God has made man *"a little lower than the angels...."*

The original Hebrew word for "angels" in this text is **Elohim.** It is the first biblical name for **God himself,** for Genesis 1:1 records: "In the beginning **Elohim created the heavens and the earth."**

Not only has God made man like himself in terms of spiritual potential, but God has made man **just a little lower than the Creator. Man is not just a little higher than the monkeys. He's just a bit below God!** Think of the implications of such a truth.

How can man relate to God?

In 1 John 1:3, we read that *"...our fellowship is with the Father and with His Son Jesus Christ."*

Unlike any of the animals, man is able to enjoy a personal and meaningful relationship with God. And it is clear from the Bible that the only way to achieve that fellowship is through the Master of creation and life itself, Jesus Christ.

In what realm can spirit-man think?

In Philippians 4:8, we are given some guidelines on how to operate our minds. Whatever things are true, honest, just, pure, lovely, good, virtuous, and praise-worthy; these are the kinds of abstract thoughts available to man. No animal is capable of thinking in these realms.

Only Man Appreciates Beauty

In Psalm 27:4, we are told to *"Behold the beauty of the Lord."* Indeed every corner of the infinite creation is filled with the Creator's endless treasure of beauty.

When we humans go out and explore the beautiful world God made, we alone can truly draw deep inspiration from it. Our innermost feelings are touched by the breathtaking majesty of an autumn sunset. We can thrill to the crescendo of a symphonic orchestra. Even the shimmering iridescence of a tiny tropical fish can stir our sense of awe.

Show any of these marvels to a hog or a horse. They'll just snort and ignore your discovery as they go back to their feeding.

What is spirit-man's authority?

In Genesis 1:26, God reveals His amazing intention to give man the total dominion over every other creature on the earth. Psalm 8:6 and Hebrews 2:8 reflect this too.

No other creature in heaven or on earth was given such supreme authority. God has left nothing that is not subject to man.

Man was designed to be a ruler, not an animal. He was created to fellowship with God, not to wander far from Him. Think about that as you review the attitudes and behaviors of those who spend their lives picking through the dust to try to find a bone that might be their cherished missing link... a link they will never find because man has always been truly human!

Why Is This Subject the Most Important Battleground of History?

Despite all the intellectually vain imaginations about mankind's origin, the clear evidence reflects the biblical truth that man has always been man from the beginning.

At the opening of man's history, the Deceiver of all time managed to skillfully convince our first parents that they should listen to him and disobey God. Satan is still doing that today.

One of the most subtle and wicked deceptions Satan has used is the lie that men are evolving and getting better along with the rest of the natural world. This unbelievable dogma is epidemic today worldwide. So, no matter how brilliantly educated a child becomes, when he is influenced by this deception, his created human dignity is robbed from him. Taking advantage of their self-centered willfulness, Satan convinces men to manifest their natural animal appetites. Yet, not even the basest animals behave with the perversion of those so deceived.

The tragic result of evolutionary thinking is exemplified in the story of Ota Benga. Darwin's book, *The Descent of Man*, prompted some anti-Christian intellectuals to seek fossils proving man rose from apes. Others believed that "half-man half-ape" creatures still flourished in remote parts of the world. In the early 20th century, such beliefs led to wanton killing of Australian Aborigines and the sad story of an African Pygmy man named Ota Benga. An evolutionist

researcher captured Ota Benga in 1904 in the Congo. In his own language, his name meant "friend." He had a wife and two children. Chained and caged like an animal, he was taken to the St. Louis World Fair where evolutionist scientists exhibited him as "the closest transitional link to man." Two years later, they took him to the Bronx Zoo in New York and displayed him as one of the "ancient ancestors of man" along with a few chimpanzees, a gorilla named Dinah, and an orangutan called Dohung. The zoo's evolutionist director made speeches boasting that this exceptional "transitional form" was in his zoo. They caged Ota Benga as if he were an ordinary animal. Demoralized and distraught by his treatment, Ota Benga eventually committed suicide.[15]

The molecule-to-man hoax of evolution has been called the "grand delusion." It is vitally important to realize that the ongoing search for fossil evidence has been fueled by an absurd idea. That idea is the fabrication that creatures can miraculously produce offspring with new improved genetic information. This idea was not generated by scientific discoveries. In backward pagan culture, such a myth might be expected to come out of pure ignorance. But in our modern culture (even our "post-modern" culture as some call it now) the myth is required by self-serving experts who embrace the prejudice that the God of Genesis could not have created according to His Word.

The concept of a paradise lost – a golden age – is a part of history that has almost been erased from modern education by teachers of evolution. Even the Greek writer, Hesiod, wrote in the eighth century B.C. of a primeval time when man was innocent and in perfect harmony with the Creator. His story explained that the curiosity of the first created woman brought on the disease and subsequent misery of the entire world. Her name was Pandora. That account of man's fall from paradise is very similar to the Hebrew version in the beginning chapters of Genesis. Could there have been a commonly known oral tradition throughout the ancient world before Moses transcribed our familiar biblical history in 1500 B.C.? The root cause of such a drama could only have been the fact that something like it really happened. And we shall see that there are even more historical facts to confirm our confidence of its truth.[16] The apostle Paul warns:

"Beware lest any man spoil you [or takes you captive] through philosophy and vain deceit [empty deception], after [or according to] the tradition of men, after the rudiments [or the elementary principles] of the world [i.e. the ungodly world system], and not after [or according to] Christ" (Colossians 2:8).

A captive is a slave. Throughout history the victorious armies became the brutal masters of the nations they spoiled. Such slavery is demeaning, depressing, and ultimately deadly.

Is it possible that philosophies can also have such painful results? Can men's traditions lead us into a deadly trap? God reveals more through the prophet Isaiah (in chapter 5, verses 12 and 13) saying:

> **...they regard not [don't pay attention to] the work [the deeds] of the LORD, neither [do they] consider the operation [the works] of his hands. Therefore my people are gone into captivity [made slaves] because they have no knowledge [a severe lack of knowledge]....**

The Creator and true science have the same objective! What is that? It is to expose deception and reveal the truth that will always lead to the greatest liberty. That is why God says:

Prove [test] all things. Hold fast that which is good.
1 Thessalonians 5:21

Don't ever be afraid to put challenging ideas to the test. But don't do it with the arrogance of trying to prove that your preconceptions are right. The more you test things against God's Word the more you realize the truth of Solomon's words:

> **Every word of God is pure: he is a shield unto them that put their trust in Him. Add thou not unto his words, lest he reprove thee, and thou be found a liar.**
>
> *Proverbs 30:5-6*

"For whatsoever is born of God overcomes the world: and this is the victory that overcomes the world, even our faith. Who is he that overcomes the world, but he that believeth that Jesus is the Son of God?" *1 John 5:4-5.*

Is There Evidence of Humans Buried by the Great Flood?

"And God said unto Noah, 'The end of all flesh is come before me; for the earth is filled with violence through them; and behold, I will destroy them with the earth.'"

Genesis 6:13

A Challenge to Prove All Things

You might encounter some people so stubborn in their rejection of the Bible and God that you expect them to say: "Don't confuse me with the facts, my mind is already made up!" Jesus alerted us to such a resistant attitude when He said **"If they hear not Moses and the prophets, neither will they be persuaded, though one rose from the dead."** (Luke 16:31)

However, many others are just confused by the public "disinformation" campaign that has blacked out facts that support a biblical view of creation, the Flood, dinosaurs and ancient man.

THINK! How would people's thinking be affected by the discovery of truly human bones or sophisticated man-made artifacts found deep in the sedimentary rock layers of earth?

Evolutionists them-selves have stated that if only one such verified discovery were made, the whole gradualistic scheme of evolution would be in ruins.[17]

Human Skeletons Found in Dinosaur Rock Layers

In 1971, a rock collector spotted some bones recently exposed by a quarry bulldozer in hard sandstone near Moab, Utah. Requesting the excavators to pause their earthmoving operation, he brought a university anthropologist, a journalist and photographer to the site. The lower halves of two human skeletons were removed and taken to the university for further study. The rock formation was confirmed to be the same "100 million-year-old sandstone" containing dinosaur bones not far away at the famous Dinosaur National Monument near Vernal, Utah.

Strangely, the bones were never subjected to the technical analysis expected. No scientific report was released to the press, and the discoverer had to reclaim his fossils.[18]

Think ! Did the location of the human bones scare off the evolutionists and prevent a careful investigation by "establishment" scientists?

The bulldozer driver points out the original location where he found the first bones of Malachite Man in 1971. He expresses certainty that there were no tunnels or cracks in the extremely hard overlying layers of rock.

Malachite Man

In 1990, an independent team of researchers, including Dr. Don Patton, excavated further and found more. Thanks to their work, it's now clear that skeletons of ten modern humans were buried under fifty-eight feet of Dakota Sandstone, in an area spanning about 50 by 100 feet. This rock formation is called Lower Cretaceous and is supposedly 140 million years old. At least four of the ten bodies are female. One is an infant. Some of the bones are articulated. Some are not, appearing to have been washed into place. No obvious tools or artifacts were found associated with the bones.[19]

The bones are partially replaced with malachite (a green mineral) and turquoise. The name "Malachite Man" is appropriate.

This perfectly modern human jaw bone and teeth are completely replaced by turquoise.

1971 Excavation: These bones, from two different individuals, a male and a female, were among the first bones found at the site.

1971 Excavation: A close up of these perfectly modern human bones.

Some insist this is a mass grave. Think about that! Who would dig a grave up to 54 feet deep through extremely hard sandstone layers? The modern mining operation was halted in the 1970s because the sandstone was so hard it was wearing out the bulldozers.

These humans and dinosaurs appear to have been buried by the same catastrophic flood! Humans and dinosaurs must have lived at the same time!

It seems obvious that these 10 men, women and children, were buried rapidly by some catastrophe, like a flood. Articulated skeletons indicate rapid burial.[20] Some argue that these people were mining in a cave, when the ceiling collapsed on them. However there are no signs of tunnels. Women and small children would not likely be included in a mining operation. No tools have been found and there are no crushed bones as you would expect if a mine caved in.

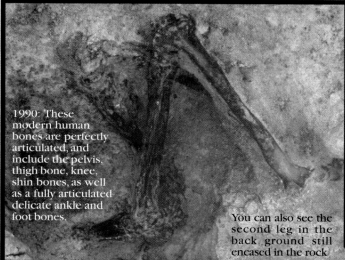

1990: These modern human bones are perfectly articulated, and include the pelvis, thigh bone, knee, shin bones, as well as a fully articulated delicate ankle and foot bones.

You can also see the second leg in the back ground still encased in the rock

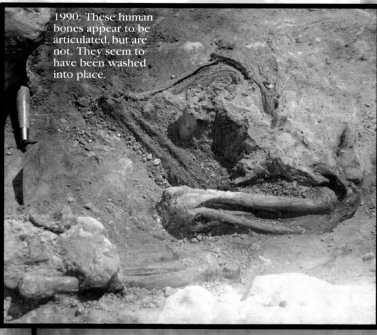

1990: These human bones appear to be articulated, but are not. They seem to have been washed into place.

Discovering the Dinosaurs

There probably isn't another subject more fascinating to young and old alike than the dinosaurs. You probably know some boys about ten years old who just love to play with and even collect models of dinosaurs. I collected them when I was a kid myself (way back in the dark ages).

One of the finest displays of life-like replicas of dinosaurs is in Canada. The pre-historic park at the Calgary Zoo in Alberta is a great experience with extinct giant animals.

Coming to the entrance of the park, I noticed an important sign designed to set the stage for our visit.

THE DINOSAURS REIGNED FOR 140 MILLION YEARS. THEY MYSTERIOUSLY BECAME EXTINCT ABOUT 60 MILLION YEARS BEFORE MAN APPEARED ON EARTH.

Does that sound familiar? The concept is typical in all introductions to dinosaurs. So it's not surprising that when we think of dinosaurs most of us have been programmed to believe it was millions of years before man ever arrived on the scene. But why?

Just because we don't see huge lumbering beasts like triceratops running around in people's back yards today, does that mean they were never seen by people at any time in earth's history?

The word "dinosaur" was invented in about 1840 after an English doctor's wife found a curious fossil that sparked the modern fascination with these powerful beasts. Over the years since then, some 200 different varieties of dinosaur skeletal parts have been dug from the earth's sedimentary rock layers.

Triceratops (try-SAIR-uh-tops)

His name means "three-spiked head."

His overall length was up to 25 feet (about as long as a large delivery truck!)

• Up to 10 feet high at the ridge of his back

• From horned collar to snout, the skull was about 8 feet long.

• Two massive horns over the eyes, 40 inches long and almost a foot wide at the base

• A strikingly handsome and powerful creature that thrived in great numbers before they were caught in a tremendous flood

Tyrannosaurus (tie-RAN-uh-SAWR-us)

His name means "tyrant lizard."

The terrifying-looking tyrannosaurus, we're told, is supposed to have suddenly died out some 60 million years ago. Imagine how much food he had to eat daily just to keep going.

He is up to 50 feet long (as long as a railroad boxcar).

As high as 18 feet, he could rest his chin on the roof peak of an average house.

- Weighed up to 20,000 pounds
- Skull length measured over 4 feet.
- Claws on hind feet were up to 8 inches long.
- Teeth like daggers were up to 6 inches long.
- Small forelimbs seem mysteriously useless to the experts.

Though many feel this was the fiercest animal ever, recent studies show it could not run very fast or grasp any prey, and was probably more docile than Hollywood has fabricated it to seem.

Dr. Duane Gish's excellent beginner book, *Dinosaurs by Design*, is a good way to open a deeper and more realistic historical understanding about dinosaurs. Be sure to add it to your library.

The Mysterious Vanished Giants

A variety of strange giant creatures once flourished all over the earth.
We're told that the "Age of Reptiles" lasted over 100 million years.
Then they all mysteriously vanished into extinction.
All we find now are the bones of these once-great titans that make present-day animals
appear like dwarfs by contrast. On every continent of our planet, men have dug their
petrified skeletons from the rock layers of the earth. Over 200 different kinds of
dinosaur have been named so far. After the bones are stabilized and the missing ones
filled in, complete skeletons or plastic castings of them are assembled and
displayed in museums all around the world.

Dimetrodon (die-MEE-tro-don)

- Unusual for its sail-like fin, the purpose of which is a mystery to the experts

- Up to 11 feet long and 6 feet high at the top of the fin

- Weighed over 650 pounds

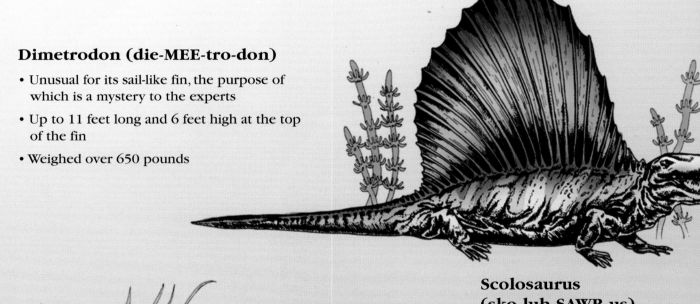

Scolosaurus (sko-luh-SAWR-us)

- This low-to-the-ground "living tank" must have been an invincible vegetarian with all his armor.

- Up to 18 feet long and 8 feet across the mid-section.

- Covered with spike-studded armor, the knobs stuck out 4 to 6 inches.

- His bony-knobbed tail wielded two spikes to ward off unwelcome antagonists.

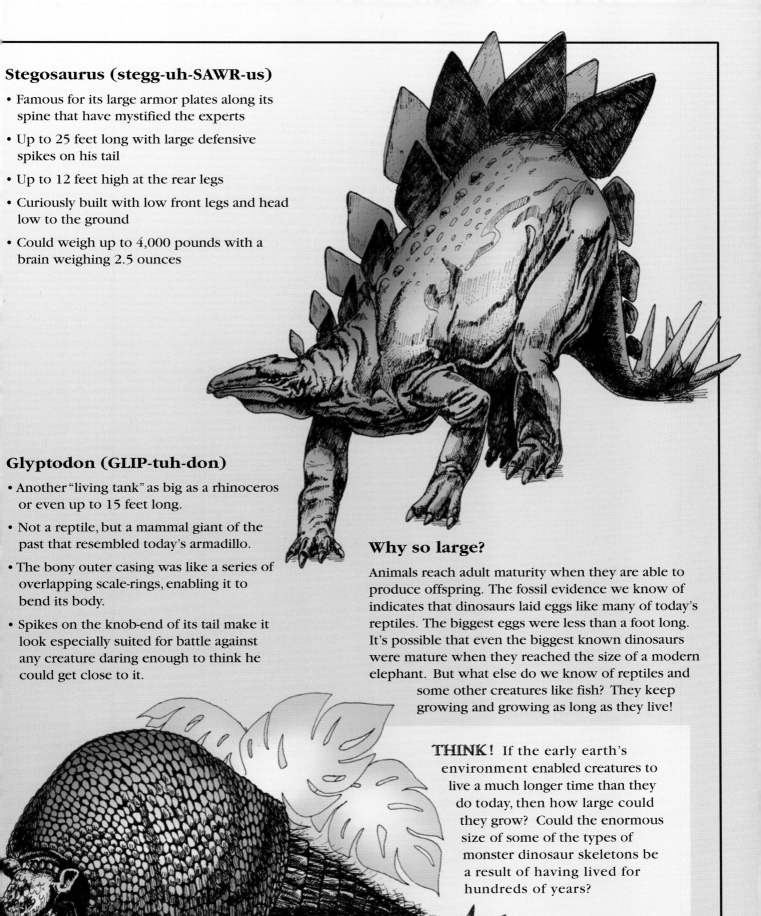

Stegosaurus (stegg-uh-SAWR-us)

- Famous for its large armor plates along its spine that have mystified the experts
- Up to 25 feet long with large defensive spikes on his tail
- Up to 12 feet high at the rear legs
- Curiously built with low front legs and head low to the ground
- Could weigh up to 4,000 pounds with a brain weighing 2.5 ounces

Glyptodon (GLIP-tuh-don)

- Another "living tank" as big as a rhinoceros or even up to 15 feet long.
- Not a reptile, but a mammal giant of the past that resembled today's armadillo.
- The bony outer casing was like a series of overlapping scale-rings, enabling it to bend its body.
- Spikes on the knob-end of its tail make it look especially suited for battle against any creature daring enough to think he could get close to it.

Why so large?

Animals reach adult maturity when they are able to produce offspring. The fossil evidence we know of indicates that dinosaurs laid eggs like many of today's reptiles. The biggest eggs were less than a foot long. It's possible that even the biggest known dinosaurs were mature when they reached the size of a modern elephant. But what else do we know of reptiles and some other creatures like fish? They keep growing and growing as long as they live!

THINK! If the early earth's environment enabled creatures to live a much longer time than they do today, then how large could they grow? Could the enormous size of some of the types of monster dinosaur skeletons be a result of having lived for hundreds of years?

What Now-Extinct Animals Lived with Man in the Past?

THINK! Knowing the evidence we discussed earlier (that life on the earth and even the planet itself could not have existed millions of years ago), is it possible that another "mystery" of creation may be resolved? How do the dinosaurs fit with the Bible?

Could dinosaurs have been seen in the past by living humans?

With all the dinosaur bones buried all around the world... and with the Genesis concept that all God's creatures were placed under the dominion of Adam... does the Holy Bible give us any hint that dinosaurs existed with humans? Keep in mind that the word **"dinosaur"** was first invented in England in about 1840 by the eminent anatomist, Richard Owen, who began studying the newly discovered fossils of these mysterious giants. Of course the Chinese people had been digging up **"dragon"** bones and selling them for years to make magical potions. Their historic concept of the memorable existence of living giant reptiles would have been no surprise to the man in the ancient account we'll read next.

4,000 Years Ago

The ancient record of a man named Job is found in the Hebrew Bible. Because of the details in this story we know that it occurred about 2,000 years before Christ, during the time when Abraham was alive. The earth must have changed a lot since then because Job lived when Arabia was good for pasturing herds and flocks of livestock. Even rivers and ice were reported in a land that today is a desert with neither of these common natural conditions.

The fascinating story of Job is one of the oldest pieces of literature on earth. It was recorded just a few hundred years after the flood of Noah's time.

In the 40th chapter of Job's account we see the record of the Creator himself speaking to Job. He is drawing Job's attention to one of the wonders of the creation. Let's read it...

From Job Chapter 40

Behold now Behemoth, which I made with thee... (verse 15).

God is saying to Job: "Look at this creature.... I created him along with you." Do you get the feeling God is continuing another segment in Job's divine nature study, begun in chapter 38? No one today knows absolutely for sure what the name "behemoth" means. The traditional understanding to the Hebrew mind was that it was a large beast. Some reference Bibles show a marginal note explaining that behemoth is an elephant or a hippopotamus. Well, is that really what Job is seeing? Let's investigate further.

...he eats grass as an ox (verse 15).

The behemoth must have been a vegetarian, and much larger than an ox.

Lo now, his strength is in his loins [or hips]... (verse 16).

This critter must have had powerful legs. Maybe it could be an elephant?

Elephant

...*his force is in the navel of his belly* (verse 16).

He apparently had a massive mid-section and could really throw his weight around. Now, if that was all we had to go on we could figure Job might have been looking at a hippo, but there's more information yet.

He moves his tail like a cedar [tree]... (verse 17).

Now we've got a problem. Have you noticed what's distinctive about a moving cedar tree (when the wind blows)? It's an enormous swaying pillar, isn't it? Now, have you ever seen the tail of an elephant or a hippo? It doesn't seem that a cedar tree is a very appropriate analogy does it? But what other large animal do we know of that lived on the earth with Job?

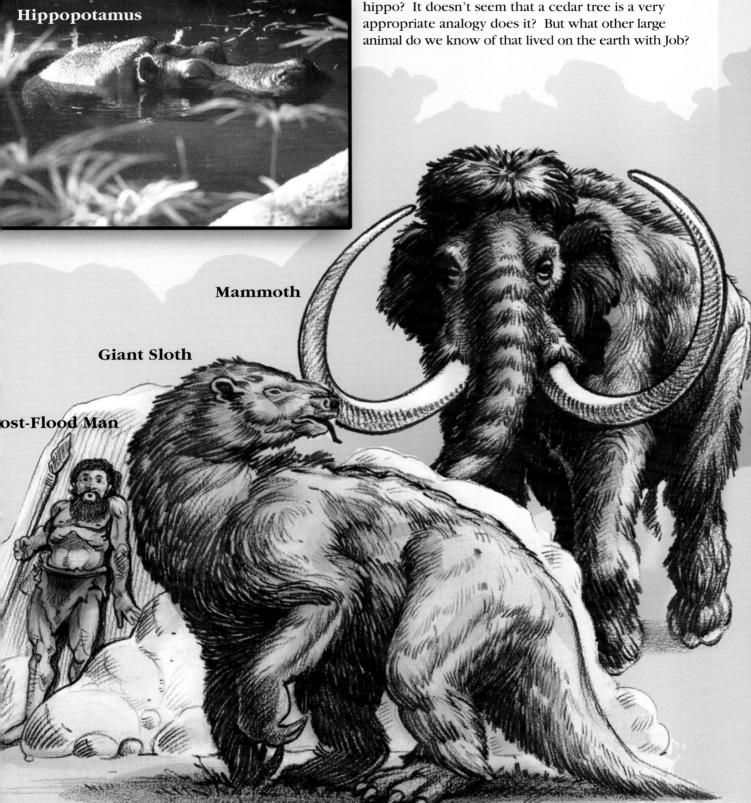

Hippopotamus

Mammoth

Giant Sloth

ost-Flood Man

Could Job Have Seen These?

Behold Now Behemoth...

Look at the tail of a creature named apatosaurus (which used to be called brontosaurus). You wouldn't want to be anywhere close by. Apparently they didn't just drag their tails along either, but could swish them around like a gigantic whip. We know that giants like these really did roam on our planet because we find their huge bones buried in different parts of the world.

The apatosaurus grew to some amazing sizes after living possibly hundreds of years. Imagine the tonnage of vegetable matter he could pack away in a day.

Can you imagine what a mighty demonstration of power would be exhibited by a living diplodocus? He truly would have been a good example of the awesome power of God displayed in the animal kingdom wouldn't he?

And just what does it go on to say in the 40th chapter of Job about this beast?

"He is the chief of the ways of God!"

His bones are as strong pieces of brass; his bones are like bars of iron (verse 18).

He is the chief of the ways of God... (verse 19).

Whatever this creature is, it's described as a supreme demonstration of God's power. Has there ever been an animal on the planet that fit this kind of superlative description? Could it possibly be a dinosaur? Our problem is that most of our generation have been brainwashed to believe that all the dinosaurs died out millions of years ago, but that leads us to...

The Next Question – Is there any hard evidence that the gigantic animals of the past were ever seen alive by humans?

Since animals and man were both created on Day 6 of creation, we would expect that some evidence would have been preserved in the vast fossil remains generated by the Flood.

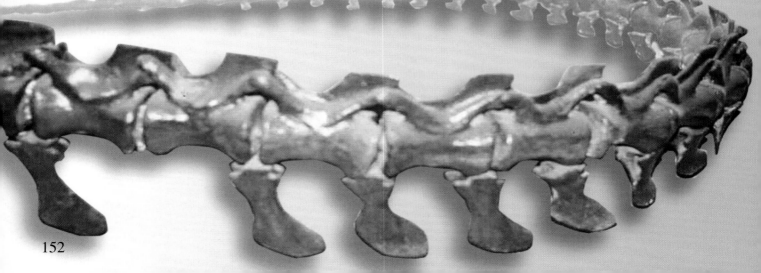

Footprints Frozen in Stone

Just an hour's drive south of Fort Worth, Texas, there is a river named Paluxy that winds through the rocky wooded countryside near the quiet rural town of Glen Rose. The river cuts through dozens of layers of hard limestone rock that has drawn curious attention since the 1930s. Why? Because layer upon layer of these calcium-rich, cement-like strata are filled with thousands of well-preserved footprints of various types of dinosaurs — they are literally footprints frozen in stone!

These cream-colored, fossil-filled layers are called "Cretaceous" (meaning "chalky") by geologists who associate them with similar rocks found in other parts of the world. They are assigned to the age of the dinosaurs. In this region, the limestone strata cover hundreds of thousands of square miles from North Texas to Mexico, Louisiana and even under the Gulf of Mexico. They are hundreds of feet thick.

Limestone is the raw material for cement. It never was ordinary mud. It had to harden very quickly to preserve the fossils and foot tracks found here. Imagine this vast plain with no plant life in sight. Many dinosaurs and other animals left their tracks here as they wandered in search of food. Only a few sparse fossils of tree branches are found coalified among the narrow bands of clay between the hard limestone layers. This is not evidence of a normal ecosystem. The aftermath of a catastrophic flood devastation is unmistakable.

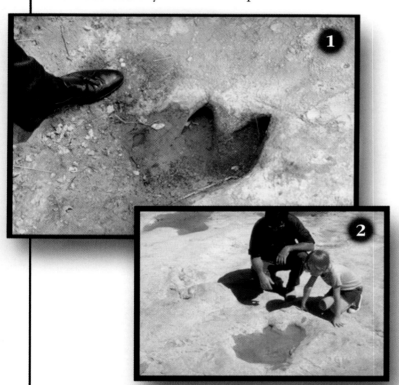

1, 2. Three-toed dinosaur tracks are easy to spot in the summer when the river level drops exposing many cement-hard platforms of flat limestone.

3. On the higher ledges beside the river, expedition participants help clear the rock and expose more tracks.

4. Since the 1930s, local ranchers have found and excavated human-looking tracks. Some of them were giant in size.

Three-toed dinosaurs similar to *T. rex* left their footprints here. So did the gigantic apatosaurus. But other creatures left their marks too. Fossil tracks of a large cat and what seems to be a bear track with distinct claw marks have also been found.

Well-documented reports from local residents claim there are even human tracks impressed into the hard limestone. This has created quite a controversy, especially by people who haven't seen the tracks and who are convinced of the evolutionist story that the dinosaurs died out millions of years before men walked on the earth.

What could have made these tracks?

For years the skeptics passed off these mysterious human-looking tracks as the creative work of ancient Indian carvers. Why did they say that? They never accused Indians of carving the dinosaur tracks!

To disprove the carving theory, filmmaker Stan Taylor began excavations in 1969 to locate more tracks under previously undisturbed rock layers. At first only a few human-looking tracks in left-right sequence were visible. So a bulldozer was used to remove the bank where the trail led. Nine more very humanlike tracks were found. This did away with the idea that the tracks were carved. Some scientists who examined the new findings admitted they looked human. Others were doubtful, although it was clear that the tracks weren't made by dinosaurs.

Later excavations extended the trail to a total of fifteen tracks in a consistent right-left pattern. The entire sequence was nicely exposed when a drought in 1999 dried up the riverbed. A trail of three-toed dinosaur tracks can be seen crossing at an angle of about 30 degrees.

5. After being washed by the river for years, isn't it amazing that you can even find evidence for the basic outlines of human feet like this?

6. A lifelike model of a dinosaur in the nearby state park shows how the dinosaur foot is the only explanation for these many giant 3-toed tracks.

7. This clear paw print of a giant cat is almost a foot across. What is a cat track doing in the Cretaceous layers with the dinosaur tracks?

8. The Taylor Trail, where fifteen human tracks intersect the tracks of a dinosaur!

9. Stan Taylor draws attention to a set of clear footprints.

A Controversy That Provokes Professional Vandalism?

What do you think would be the repercussions of finding true fossilized human footprints in the same rock as dinosaur footprints? Even evolutionists have admitted that such a discovery would completely overturn the prevailing theory of evolution and its entrenched concept of millions of years. To some people, the possibility is so objectionable that it cannot even be allowed.

Back in the Depression days of the 1930s the human-looking tracks were a curiosity that prompted some of the local ranchers to take the time to chisel some of them out of the bedrock. Some were sold to collectors. Accusations have been made that even fake tracks were made and sold, but I've not seen any. I first visited the Paluxy River track site in 1975 and asked the ranger on duty at the state park about the human tracks. I was amazed at his answer. He said he didn't know about them, and obviously he didn't want to discuss it. Everybody in town knew, but he was clueless... or was he ordered not to talk about it?

In 1988, as the director of the Creation Resource Foundation, I began co-hosting a series of annual research team investigations of the mysterious tracks. Hundreds of people accompanied us over the years to see the evidence first hand. Much careful work was done to search for, expose and document the tracks. Everyone, even skeptics were amazed by the convincing evidence when seeing and feeling the rocks in person. The real eye-opener was encountering the deceiving behavior of the self-appointed anti-Christian skeptic squad who showed up uninvited to distract and intimidate our guests. They were not there to discover... only to confuse. The reality of spiritual warfare over origins became more evident than ever.

1. This cast made by Stan Taylor in 1970 appears to be the +1 right foot impression of the 15-track sequence called the Taylor Trail. The tracks are consistently 11.5" long with appropriate shape details for each step.

2. Here is the +1 track in 1988, showing 18 years of erosion. See the large chunk lost from the left side. The same general shape is clear.

3. After the August 1989 presentation by Don Patton, here is how the +1 track appeared, having been badly battered.

Over the years, the tracks are changed by erosion and ???

On August 12, 1989, Mr. Don Patton gave a slide-illustrated lecture at a creation conference in Dayton, Tennessee. He presented compelling evidence that human and dinosaur tracks were together at the Taylor Trail, including the first two pictures on the previous page. Two well-known evolutionists were present. At least one was conspicuously disturbed by the presentation. Both flew to Dallas the next morning and went immediately to the Paluxy River. It is reliably reported that they were in the riverbed that afternoon with an "iron bar." Three days earlier the fossil tracks were observed looking as they normally had been. Three days after they were in the river, the +1 track and others were observed severely damaged. Clear photos were not possible until the river level went down a few months later.

A local Glen Rose rancher named Jim Ryals managed to chisel out a slab of limestone from the Paluxy River in the 1930s with a beautiful human track in it. That track was on display for years in the courtyard of Dr. Cook's medical clinic in Cleburne, Texas. When Stan Taylor tried to chase it down, it was missing. The hole left in the river's bedrock is still there right in front of a fairly good track that has come to be known as the "Ryals Track."

The toe section of the Ryals Track (imprinted toe-holes) extends under an overhang (of hardened limestone), apparently formed when the foot was pulled out of the calcareous mud before it set hard. This same kind of structure can be seen in some of the dinosaur tracks found in the same layer of limestone. This is one of three tracks that were featured by Mr. Patton at the 1989 creation conference in Dayton, Tennessee. It was destroyed the next day.

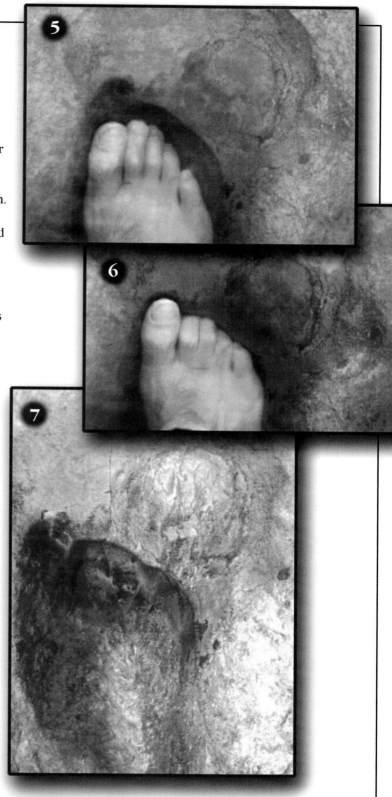

4. With the water removed, the "Ryals Track" even shows toe scratch marks on top of the toe area where the toes dragged as they came out of the wet cement before it had hardened to stone.

5. I remember fitting my foot snugly under the overhang of the Ryals Track in July of 1989.

6. See how nicely a man's foot fits into the Ryals Track?

7. Closeup of the Ryals Track (before it was destroyed) showing prominent toe features.

The Paluxy Tracks

The Case Is Still Open

In 1986, news stories circulated that creation scientists had finally given up on the Paluxy tracks. Curious dinosaur foot-shaped rust stains began showing up that supposedly disqualified the elongated tracks from being human. Some notable authors who had never personally witnessed the tracks were influenced by some inconclusive reports. Others who didn't have first-hand knowledge of the whole gamut of evidence took a wait-and-see position. But the field re-examination and more excavations continued.

In 1992, a clarification was made of the +6 track in the Taylor Trail. The Taylor Trail aroused special interest because the "human" tracks in left-right-left progression were imprinted inside a series of obvious dinosaur tracks. Some skeptics insisted the human look was created coincidentally by odd structures on the bottom of the dinosaur feet that made them. But now the rock revealed that this +6 track of human form was **beside the next dinosaur track** in the series rather than inside it.

Excavations have continued to expose more tracks showing both human-like and dinosaur tracks together. A Japanese team of scientists uncovered tracks in 1996 that led one of them to exclaim that it was almost enough to convince him to become a creationist.

The Burdick Track

Since the 1930s, Dr. Clifford Burdick had possession of a clear five-toed giant human-looking track that gave the skeptics fits. Because the track came from the same limestone of the Paluxy River area where the dinosaur tracks are found, evolutionists had to insist it was a carving.

1. The +6 track (left foot) of the Taylor Trail was uncovered under six feet of alternating layers of clay and limestone. The surrounding "mud push up" as well as the toe depressions, human length and width proportions, and the proper distance from the previous right track all weighed heavily in concluding this vindication of the authenticity of the tracks.

2. A lower sideview of the +6 track while it is under a slight amount of water accents the relative size, shape and positioning just behind and to the left of the imperfect dinosaur track.

H-T
+6
L

The clear and perfect placing of the five toes, the perimeter push-up ridge, the raised middle from suction of the lifting foot, as well as the complete lack of tooling marks does not leave many options to interpret the origin of this track. Skeptics insisted that slicing the fossil would reveal no compaction of the crystallized stone under the track and thus it would have to be a fake. A carving would randomly cut across the internal rock structures. However, if compaction structures follow the contours of the impression, the carving theory would be falsified.

When the slicing was done in 1989, internal compaction structures were found to dramatically conform to the shape of the heel impression and the great toe impression. This demonstrated that the Burdick track really is an original impression of a foot with five toes and elongated shape found in a limestone layer that is well known for dinosaur tracks.

The extra wide span of the Burdick track makes us question what kind of foot made the print. However, tests made of students running through wet cement showed the same peculiar characteristics: pronounced toe impressions, raised centers, wide phalanges section, and narrow heels. You might be surprised at all the strange shapes you get from people running through stiff mud.

After a detailed scientific analysis assigned by the Bible Science Association of Minneapolis, engineers Robert Helfinstine and Jerry Roth published a careful technical assessment of the Paluxy River tracks in 1994 titled *Texas Tracks and Artifacts*. Reading this work will help you understand why those wholly committed to evolution will do everything they can to discredit it.

4. The side view of the cross cut through the big toe of the Burdick track shows an outstanding color variation that conforms to the depression above. This could not result from a carving.

5. The 15-inch-long Burdick track represents a fellow that was probably at least 7 feet tall. The carving allegation is now falsified!

3. The side view of the cross cut through the heel of the Burdick track shows variations in color indicating compaction in conformity with the contour of the depression above it.

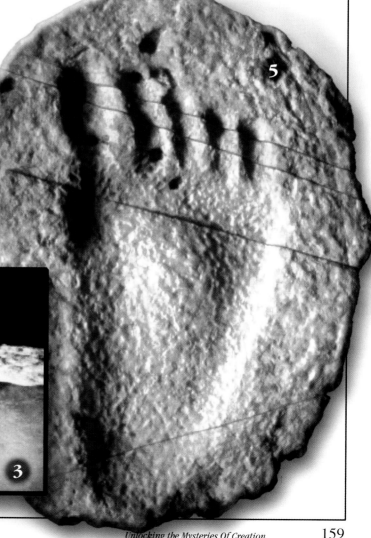

More "Problematica" for the Skeptics

Many important fossil discoveries clearly demolish evolutionary assumptions. The public doesn't even know they exist. They come as a big surprise to most of us when we hear about them. Do you suppose that's because we've been so effectively shielded from discovering what some have called "forbidden archeology?"

WATCH OUT
CONTROVERSY AHEAD

A Human Fossil Finger?

When fossilized creatures and plants are replaced by mineralized groundwater they are called "petrified." Petrified wood is often so well preserved that the cell structure and rings of a tree branch are clearly seen. Petrified fish are often so well preserved that their scales and fins are easily seen. While petrified bones are common, the impressions of dinosaur skin have also been found. Fleshy body parts are rare but even petrified worms are found in the strata where an amazing fossil was discovered.

Its remarkable! It looks like a female human finger – completely mineralized! It was found in Cretaceous rock near Glen Rose, Texas. The fingernail, cuticle and shape are easy to identify. Is it evidence of a human buried in the Flood? The absence of knuckle skin folds make you wonder if its just an odd piece of limestone, but a bloated body would explain those missing wrinkles. It was cut in two places to see any internal structure.

The cross section reveals anatomical features **highly unlikely** for just a finger-shaped rock: skin, central bone, flesh, and marrow. Another clue on the cross section is the pair of brown dots one above and one below what could be the central bone. Dr. Dale Peterson, M.D. identified these as tendons in the CT scan of the fossil. Notice the larger dot is on the underside of the "finger," stronger for grasping things. The narrower tendon on top is used to spread the finger, requiring less strength and size. At one display of it, sixteen physicians all agreed that **the features of this odd fossil match a human finger**. Many reject the evidence because they believe human fossils cannot exist in dinosaur strata. Some even insist the fleshy features of a finger could not be preserved. Their preconceptions seem to over-rule the technical analysis of what the object appears to be.[21]

1. Before the fossil was sliced, it was easy to identify as a human finger.

2. Two slices were made to see if it was just a plain rock inside.

3. Coloration and density differences reflect anatomical accuracy.

The Perfect Human Handprint Fossil in Cretaceous Limestone

The Cretaceous limestone of the Glen Rose formation in Texas is a fine-grained, light cream-colored rock that is very hard to break. To preserve impressions in this rock required it to be quite sticky "cement" before it hardened very quickly. Finding an impression of what looks like a human hand is truly remarkable. It was discovered in the late 1990s.

Veteran fossil excavator, museum preparator and researcher, Joe Taylor of Crosbyton, Texas examined the fossil carefully and reported that the surface texture of the palm print would be almost impossible to sculpt. When a latex mold was removed from the handprint, the fingernail from the second finger was imprinted. This area of the fossil print is recessed and impossible to carve. Joe is also an expert sculptor. He commented that the "web" of the fossil between the thumb and the index finger would also be unlikely the work of a "carver" who would not even think of such detail. [22]

4. This handprint has retained clear detail including fingernail and webbing impressions. It's hard to believe that an early human would have the knowledge, technology, or even the desire to carve a perfect handprint into an extremely hard section of limestone.

Another Too Perfect Fossil Footprint

In a remote part of New Mexico a clear human foot impression was found in 1929. Don R. Patton, one of the most experienced researchers on the Paluxy River tracks, managed to get to this top secret site after *Smithsonian* magazine published an article in July of 1992. The magazine acknowledged "what paleontologists like to call 'problematica.'" It described what appeared to be large mammal and bird tracks that "evolved long after the Permian period, yet these tracks are clearly Permian" (to an evolutionist that means the rock was deposited 250 million years ago). The five-toed track, called the Zapata track, is on a ledge of very hard limestone beside where another partial track apparently broke off many years ago. Stone-cutting saws were unable to cut through the dense hard limestone. That makes it hard to believe the theory the track was carved by some recent human. The raised material around the track and between the toes strongly suggests it is a genuine human foot impression. [23]

5. Close-up of the Zapata track shows detail too perfect for some to believe it's real, but the rock is too hard for it to be a carving.

6. Don Patton beside the Zapata track in New Mexico.

Can Ancient Reports of Real Dragons Be Taken Seriously?

Historical perspective is often missing from the halls of modern education. Much ancient knowledge has been lost due to the senseless destruction of invading armies. When the famous ancient Library of Alexandria, Egypt was destroyed, it was a huge loss to scholarship. The conquest of Mexico resulted in the burning of all but a handful of elaborate Aztec and Mayan codices (books). Since the popularization of evolution, it's not just the biblical accounts of the Flood and creation that are spurned by many intellectuals. Even historical events passed down through generations of common folk are ridiculed as myths. Seldom do modern students have the chance to hear the believable records of ancient people professing knowledge of things foreign to our experience. Many kinds of animals called dragons are real to traditional cultures worldwide. Should we spurn these stories, or rather explore them carefully?

Though dragons have almost been universally omitted from modern studies of natural history, they were considered a normal part of orthodox nature study prior to the introduction of evolutionary ideas in geology (in the late 1700s).

Leviathan?

In our modern culture, the idea of Chinese dragons is routinely dismissed as unreal mythology. But the Chinese people historically are among many cultures that knew dragons once existed. Consider the 41st chapter of that 4,000-year-old Hebrew account called Job. God is talking to Job and asking Job some questions about an amazing sea creature: **"Can you draw out leviathan with a hook?"** (verse 1). More questions reveal that this creature is practically impossible for man to restrain. The 1599 edition of the Geneva Bible conveys the impressive power of this beast with the conclusion of the passage in the last two verses of the chapter:

In the earth there is none like him; hee is made without feare; he beholdeth all hie [high] things; he is a King over all the children of pride.

Helping Job understand God's power through a created beast, do you notice some even more remarkable parts of God's description of leviathan here?

Again, reading from the Geneva Bible:

Lay thine hand upon him; remember the battell, and doe no more so.... No one is so fierce that dare stirre him up. Who is he then that can stand before me?... Who shall open the doores of his face? His teeth are fearefull round about. The majesty of his scales is like strong shields and are sure sealed.... His sneisings make the light to shine, and his eyes are like the eyelids of the morning. **Out of his nostrils commeth out smoke, as out of a boyling pot or caldron. His breath maketh the coales burne; for a flame goeth out of his mouth....** The mightie are afraid of his majestie, and for feare they faint in themselves.... He esteemeth yron [iron] as straw, and brasse as rotten wood.... He maketh the depth [water of the sea] to boile like a pot....

This is Holy Scripture, not just some fanciful myth! But before you feel embarrassed or critical that the Bible reports this as clear narrative, consider these marvels of creation.

Fireflies – If they weren't so common, the idea of flying light bulbs would belong to Fantasyland.

Bombardier Beetles – When you first learned about bugs that literally have a "blast" as hot as boiling water to resist their natural enemies you probably were as amazed as the rest of us... Right?

Electric Eels – 600 volts of electricity coming from an animal is enough to make anyone want to get out of the way. If it weren't a known fact you probably wouldn't believe it.

Let's face it. If we dug up a skeleton of a skunk we couldn't know that it produced an odor that drives you far away. If you dug up fossilized porcupine bones you wouldn't know about its effective defense system. **The ability of a created beast to breathe fire is not biologically impossible!**

In light of Job's leviathan, have you ever wondered where the name for the military squadron called "dragoons" originated? In the early days of firearms these soldiers were equipped with what they called a "dragon," a short musket spouting fire like a dragon when it was discharged. The head of a dragon was carved on the muzzle of the weapon. [24]

Winged Serpents?

The famous Greek philosopher, **Aristotle**, wrote that in his own time there were common reports of winged serpents in Ethiopia. [25] The Jewish historian, **Josephus**, also vouches for the reality of flying serpents at the time of the Exodus from Egypt. [26] A man named **Herodotus** is called the "Father of History" for the careful accounts he collected of ancient events. He was a Greek living in the fifth century B.C. Listen to his report:

"I once went to a certain place in Arabia, almost exactly opposite the city of Buto, to make inquiries concerning the winged serpents. On my arrival I saw the backbones and ribs of serpents in such numbers as it is impossible to describe; of the ribs there were a multitude of heaps, some great, some small, some middle-sized. The place where the bones lie is at the entrance of a narrow gorge between steep mountains, which there open upon a spacious plain communicating with the great plain of Egypt. The story goes, that with the spring the winged snakes come flying from Arabia towards Egypt, but are met in the gorge by the birds called ibises, which forbid their entrance and destroy them all. The Arabians assert, and the Egyptians also admit, that it is on account of the service thus rendered that the Egyptians hold the ibis in so much reverence." [27]

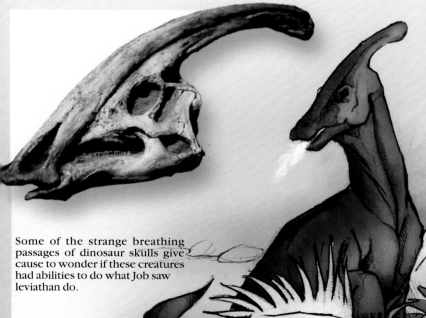

Some of the strange breathing passages of dinosaur skulls give cause to wonder if these creatures had abilities to do what Job saw leviathan do.

Where Do Dragon Legends Originate?

The First Endangered Species?

What happens when we hear the telltale rattling of a rattlesnake or hear the hissing of a water moccasin? Have you noticed that even a small alligator lizard will cause people's pulse rates to rise? Wherever people have settled, serpents and large reptiles are among the first objects of the hunt. Humans just don't seem to tolerate even miniature dragons. What do you think they would do with real ones but try to exterminate them?

Before the word "dinosaur" was invented... they called them "dragons."

The first century naturalist, Pliny, wrote:

Africa produces elephants, but it is India that produces the largest, as well as the dragon, who is perpetually at war with the elephant, and is itself of so enormous a size, as easily to envelop the elephants with its folds, and encircle them in its coils. The contest is equally fatal to both; the elephant vanquished, falls to the earth, and by its weight crushes the dragon which is entwined around it. [28]

Life in the Middle Ages had its stress.

Ancient and not-so-ancient reports of dragons are written, not as legends, but as straightforward journalism. William Caxton, England's first printer, recorded in 1484 a cow's nearly fatal encounter with a "serpent" quite different from our concept of today's snake. Engraved illustrations of medieval times show these creatures having two legs, and large mouths:

...about the marshes of Italy, within a meadow was sometime a serpent of wonderful and right marvelous greatness, right hideous and fearful. For first he had the head greater than the head of a calf. Secondly, he had a neck of the length of an ass, and his body made after the likeness of a dog. And his tail was wonderfully great, thick and long, without comparison to any other. [29]

Above– Persian horseman and dragon: This Persian painting from the 16th century (or earlier) testifies of one of many kinds of dragons living in rather recent times.

Right– Another Persian illustration of several varieties of dragon including a type of winged serpent. Note the similarities in the features of the dragons featured on this page. Keep in mind that the art on these pages comes from all around the world.

Above– Dutch engraver, Theodore DeBry, published these matter-of-fact pictures depicting what the Spanish encountered in Virginia before 1618.[30]

The Patron Saint of England

In the third century A.D., one of the most famous encounters with a dragon happened to a man who became known as Saint George, the patron saint of England. The story has become a legend, but of a real man who was honored for his courage.

There was living in a great lake a terrible dragon with breath so foul it poisoned the countryside around the lake. The local people tried to keep the beast content by feeding it two sheep a day. When they ran out of sheep they gave up their children. Finally, the king's own daughter was tied to a stake in the field waiting for the dragon to come and eat her. Luckily, George happened to be passing by. When he saw the girl tied up and crying, he stopped to investigate. She warned him to run for his life, fearing he would be killed too. But George courageously attacked the dragon, killing it with his lance. Because George gave the

glory to Christ for the victory, the princess and then the whole population were baptized as Christians.

The legend of Saint George may have been embellished in some details, but it was believed for centuries as a true account. George himself was unfortunately martyred for his faith in Jesus Christ on April 23, A.D. 303. The English crusaders revered him highly, and the English church made him the patron saint of England in 1350.

South American Dinosaurs?

Dr. Javier Cabrera and his father have collected since the 1930s over 1,100 engraved burial stones from the Ica culture of Peru. He is professor of medicine and department head at the University of Lima. A third of the stones depict the pornographic culture of Icas from A.D. 500 to 1500. Some show their idol worship. Others depict amazing accomplishments like brain surgery, confirmed by finds of scarred skulls demonstrating healed recovery of the patients.

Nearly a third of the stones show specific types of dinosaurs like triceratops, stegosaurus and pterosaurs. Several diplodocus-like dinosaurs are shown with serrated frills along their spines. This feature was unknown in modern times until 1992. Some of these stones were taken back to Spain as reported by an Indian chronicler in about 1570. Since dinosaurs like we see here were not described until the 1800s from skeletal fossils, these "snapshots" of live dinosaurs interacting with humans are unacceptable to evolutionary die-hards. [31]

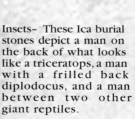

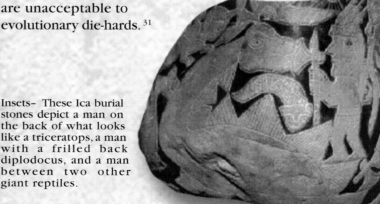

Insets– These Ica burial stones depict a man on the back of what looks like a triceratops, a man with a frilled back diplodocus, and a man between two other giant reptiles.

Could Some Dinosaurs Still Be Living?

The science of "cryptozoology" is becoming increasingly respectable with the increased awareness of many thousands of historic "sightings" of mysterious animals that can't help but bring to mind an association with the supposedly "extinct" dinosaurs of the past.

Sea Monster or Surviving Saurus?

Cadbury may be a familiar name to chocolate lovers, but cadborosaurus is a name for a giant sea serpent often sighted off the coast of British Columbia, Canada. Up to 50 feet long, these creatures have a head like a horse, a long neck and short front flippers. Its credibility increased when in 1937, a ten-foot-long juvenile specimen was found in the stomach of a sperm whale.

Respected zoologists hope to capture cadborosaurus for study.

The Mysterious Santa Cruz Carcass of 1925

Many people saw it. The press reported it widely. Washed ashore in 1925 just north of Monterey Bay, California, a giant rotting corpse was analyzed. A respected president of the Natural History Society of B.C., Canada wrote:

My examination of the monster was quite thorough. It had no teeth. Its head is large and its neck fully 20 feet long. The body is weak and the tail is only three feet in length from the end of the backbone. These facts do away with the whale theory [which had been proposed by a handful of other authorities], as the backbone of a whale is far larger than any bone in this animal... I would call it a type of plesiosaurus.

Mr. Wallace theorized the monster might have been preserved in a glacier for thousands of years, only to be released and floated to the area before washing ashore.[32]

One can't help wondering why this and other discoveries aren't more widely reported. However, sea monsters and recently living dinosaur types have not been popular among evolutionists who insist on their mythical millions of years of geologic time.

The Santa Cruz (Moore's Beach) carcass was 34 feet long, a yard wide, and had a duck-billed head the size of a barrel.

The Modern Japanese "Plesiosaur?"

Off the coast of New Zealand, a Japanese fishing crew hauled an odd catch in their nets in 1977. The giant carcass was 30 feet long and weighed about 4,000 pounds. To avoid spoiling their fish catch, the men had to dump it back in the ocean, but not before a zoologist on board made notes on the animal. Its long neck, flippers and other details were still intact enough to suggest it was some kind of saurian creature rather than a shark. [33]

Is there really a Loch Ness monster?

In the highlands of northern Scotland, people have reported a gigantic swimming "beastie" for the last 1,400 years. Some 3,000 sightings and even some vague photos seem to bear a resemblance to the plesiosaur of dinosaur times. Nessie is said to be 20 feet long with a serpent-like neck and fat body with flippers on the sides. There may be a whole family of them living in some dark submarine cavern.

Scientist Sighting More Convincing?

Natural History magazine reports the 1905 sighting by two expert naturalists aboard the private yacht *Valhalla* off the coast of Brazil. One wrote:

> I saw a large fin or frill sticking out of the water, dark sea-weed-brown in color, somewhat crinkled at the edge. It was apparently about 6 feet in length and projected from 18 inches to 2 feet from the water. I could see, under the water to the rear of the frill, the shape of a considerable body. I got my field glasses on to it and almost as soon as I had them on the frill, a great head and neck rose out of the water in front of the frill… the neck appeared about the thickness of a slight man's body, and from 7 to 8 feet was out of the water; head and neck were all about the same thickness. The head had a very turtle-like appearance, as had also the eye. [34]

From Canada's Ogopogo and Champ to lesser-known sightings in Russia and Japan, there seems to be a lot of evidence that sea serpents of one kind or another are real.

Can flying reptiles still be alive?

African explorer Frank Melland kept encountering vague rumors about a much-feared animal called *Kongamato* that lived in the Jiundu swamps in northwest Rhodesia, near the Belgian Congo. When asked what it was, the natives told him it was a bird, but more like a lizard with wings of skin like a bat's. When he showed them pictures of the pterodactyl and other animals, all immediately went for the pterodactyl, excitedly muttering "*kongamato*!"

What happened to the thunderbird?

The Tombstone Epitaph printed a news item on April 26, 1890, recalled by Dr. Duane Gish in his excellent book, *Dinosaurs By Design*. [35] The report told how two ranchers were startled by a gigantic flying reptile with huge leathery wings, a long slim body, having claws on its feet and at the joint of its wings. They said its 8-foot-long head was like an alligator's, with a mouth full of teeth and large protruding eyes. They killed it and cut off the tip of its wing for a trophy. Could this have been the last of the legendary thunderbirds? Many American Indian tribes recall the huge size and powerful flight of this strange beast that thundered when its wings flapped.

We know animals like these really did live. The problem is that we've been programmed to think they died off with the dinosaurs millions of years ago.

The 1905 sighting report off the Brazilian coast

More Terrible Lizards Still Alive?

Imagine taking an afternoon stroll in the woods and bumping face-to-face into a ten or twenty-foot-long dragon! Did you know there really is something like that living today?

Dragon Lizards of Komodo

The world's largest living lizard, according to a *National Geographic* article in December 1968, is the Komodo monitor lizard. They say it "remains something of a mystery." [36]

They are called living examples of the "prehistoric Age of Reptiles." They were totally unknown to modern man until the year 1912! Could other "dinosaurs" still be living today?

There are about a thousand terrifying lizards similar to this surviving on the remote island of Komodo in the Indian Ocean north of Australia.

Extraordinary Saurian in Bolivia

In 1883, the *Scientific American* magazine reported an "extraordinary saurian killed" in Bolivia.[37] The 39-foot-long beast is described as a factual discovery:

The legs, belly, and lower part of the throat appear defended by a kind of scale armor, and all the back is protected by a still thicker and double cuirass, starting from behind the ears of the anterior head, and continuing to the tail. The neck is long, and the belly large and almost dragging on the ground.

"Professor Gilveti, who examined the beast, thinks it is not a monster, but a member of a rare or almost lost species, as the Indians in some parts of Bolivia use small earthen vases of identical shape, and probably copied from nature.

Living Dinosaurs in the Congo?

Deep in the heart of tropical Africa lies the 55,000-square-mile Likouala swamp surrounding Lake Tele. The native Pygmy people there tell of encounters with giant dinosaur-like reptiles in modern times.

Occasional visitors to this largely unexplored region return with incredible stories of a variety of large reptiles. They all confirm that natives are absolutely honest in their reports of the "mokele-mbembe" (mo-KEL-lee-mm-BEM-bee). They say these huge animals with small heads, long necks, and long massive tails, wade in the slow meandering rivers there. They've been described as half elephant and half dragon.

Dr. Roy Mackal's carefully written book, *A Living Dinosaur? In Search of Mokele-Mbembe*, provides extensive personal narrative of expeditions to help understand the dynamics of why it is so hard to capture these elusive creatures on film.[38]

Interviews with the natives revealed astonishing things. In about 1960, some of the Pygmies managed to spear one of the creatures to death. All who ate it died soon afterward. Naturally, these simple folk are inclined to believe there is a curse associated with trying to hunt them, so they try to stay away from them. When shown pictures of various animals, living and extinct, the Pygmy people always insist their mokele-mbembe is most like the old "brontosaurus" pictured in many modern books.

The Feb. 13, 1910 edition of the *New York Herald* is one of many published reports about the giant mokele-mbembe in the Congo.

THINK! What would the discovery of a living apatosaurus do to the mythical evolutionary chart of geologic history?

Why Did Dinosaurs Become Extinct?

One of the greatest mysteries of earth science over the years has been the extinction of the dinosaurs. You've probably heard the standard evolutionary lines of how the earth was dominated by the giant reptiles for over 100 million years. Reading the popular textbooks and museum exhibits you get the idea that mammals couldn't really get started on the planet until the giant reptiles were out of the way. Then it took another 60 million years for rats to evolve to humans... more of the myth of transmutation. But what really killed off the dinosaurs?

Inadequate Theories Leave Unsolved Mysteries

The 16th century explorers to the New Wo[rld] witnessed 50-foot-long alligators being hunted [by] natives. Remember that reptiles grow as long [as] they live. Some must have lived a very long ti[me.]

One old theory insisted that "early" **mammal rodents,** no bigger than rats, went around raiding all the dinosaur nests. They ate all the eggs! The problem was that, from analyzing the fossil deposits, the dinosaurs seemed to have died off rather suddenly. And considering that the rats had to have lived alongside the dinosaurs for supposedly millions of years puts this theory in the trash heap of ideas.

Another view supposed that the **brains** of the dinosaurs, being rather tiny, eventually made them all obsolete. Were the dinosaurs too stupid to survive? They managed all right with the brains they had from their beginning. Why would anything change?

Could overweight have been the problem? Some have thought the tremendous size of the dinosaurs caused many to get "slipped disks" in their spines. This made it hard to forage for food. So they all starved to death because of **back trouble?**

In 1946, one paleontologist theorized that a worldwide heat wave of just two degrees above normal would have "baked" all the male dinosaur sex glands. With a mass

outbreak of **sterility** it wouldn't be long before every last dinosaur died without heirs.

Maybe they were **fussy eaters.** If their favorite plant food suffered severe crop failures, it had to be a sure case of starvation. Others suppose the dinosaurs didn't have enough sense to avoid **poisonous plants.**

Believe it or not, one theory suggested that the dinosaurs might very well have died out from a mass case of **constipation!**

The cover of *Time* magazine (5/6/85) featured a new theory asking: *"Did Comets Kill the Dinosaurs?"* Scientists backing this theory point to revolutionary evidence that they insist reflects a series of **cosmic catastrophes** striking the earth.

A tremendous impact supposedly raised so much dust that the sky darkened, the world's air temperatures dropped, and the cold-blooded reptiles died of frostbite.

As *Time* reported:"Whether these **catastrophic impacts** are random or cyclic remains to be seen. But if they occur at all, they could **shake the foundations of evolutionary biology…."**

The implications do present a major break with long-held evolutionary theory. But don't you think they are missing something? What about the phenomenal implications of a globe-rocking cosmic collision?

Even so, the new catastrophe theory does face facts that other extinction models ignore. As *Time* put it: " … all these fanciful concepts fail to account for the hundreds of other species that perished at the end of the Cretaceous." U.C. Berkeley physicist, Luis Alverez said, **"The problem is not what killed the dinosaurs but what killed almost all the life at the time."**

In skillful, though misleading, editorial style, *Time* concludes the article by quoting another scientist, saying: "Maybe there just wasn't enough room for them on the ark." You can imagine the jeers in the science classrooms over that one. Obviously, the editor expected a reaction. But again, maybe they're a lot closer to the truth than they would like to admit.

The scientific facts we've observed earlier point strongly to a radically altered ecosystem on this planet – a planet that once was far more tropical than it is today. Not only would many species struggle to survive in this harsher environment, but if survivors on the ark were perceived as dragons, enemies or likely trophies, you wouldn't have to think too hard to see why many of them would have been decimated by human hunters.

Fresh Dinosaur Bones?

Most people think that dinosaur bones are known only as fossils, and that this proves dinosaurs died out millions of years ago. Yet fresh, unpetrified, dinosaur bones have been found, suggesting that dinosaurs have lived more recently than many think.

In northwestern Alaska in 1961, a geologist found a bed of dinosaur bones in unpermineralized (unpetrified) condition.[39]

A young Inuit (Canadian Eskimo) who was working with scientists from Newfoundland's Memorial University in 1987 on Bylot Island found part of a lower jaw of a duckbill dinosaur. It too was in fresh condition.

The journal *Science* on December 24, 1993, (pages 2020-2023) reported on the amazing preservation of the bones of a young duckbill dinosaur found in Montana. These appeared to be fresh bones, not fossilized.

Such findings cast serious doubt on the millions of years claimed for the dinosaurs.

The Brilliant Glory of Earth's Earliest King

"And they that be wise shall shine as the brightness of the firmament [heavens], and they that turn many to righteousness as the stars for ever and ever."

Daniel 12:3

Awesome Insights from God's Word

The Bible gives us wonderful insights if we'll take the time to study and pray. As it says in Proverbs 3:5, we must put our confidence in the LORD and not in our own shallow understanding. As we do that, He will guide us into all the truth.

The Bible often reveals some amazing discoveries we would not otherwise find. As you conclude this section about "unlocking the mysteries of original man," may "the God of our Lord Jesus Christ, the Father of glory, give to you a spirit of wisdom and revelation in the knowledge of Him" (*Ephesians 1:17*).

What happened to Adam and Eve?

The event caused them to suddenly "see" their nakedness immediately after they rebelled against God's spoken will. Did they lose something? Was there something of God's nature created as part of them that suddenly vanished? How would perfect man differ from earthy man spoiled by sin?

What part does light play in the presence of God?

Jesus, the Son of Man Before God

"And as he prayed, the fashion of his countenance [face] was altered, and his raiment [garment] was **white and glistening** [flashing like lightning]."

Luke 9:29

Jesus, the Son of God Before Man

"I am the **light** of the world."

John 8:12

Moses

"…when he came down from the mount… Moses wist [knew] not that the skin of **his face shone** while he talked with him… the skin of **his face shone**; and they were afraid to come nigh him.… The skin of **Moses' face shone**: and Moses put the veil upon his face again.…"

Exodus 34:29ff

God

"God is **light,** and in Him is no darkness at all."

John 1:5

"O LORD my God, thou art very great; thou art **clothed** with honor and majesty. Who coverest thyself **with light as with a garment.**…"

Psalm 104:1-2

"…the King of kings, and Lord of lords; Who only hath immortality, **dwelling in the light** which no man can approach unto…"

1 Timothy 6:15-16

"And now men see not the bright light which is in the clouds… out of the north comes golden splendor; **around God is awesome majesty.**"

Job 37:21-22 (NASB).

"Blessed is the people that know the joyful sound: they shall walk, O LORD, in the **light** of thy countenance."

Psalm 89:15

Isn't it interesting how the inspired Word of God frequently associates the presence of God himself with awesome brilliant light? Think how physical light is so divinely unique. It is incomprehensible. Science cannot adequately come to terms with light. Its very essence is divine. The starry lights of the heavens declare the very glory of God with subtle, yet profound symbolism.

When God spoke at the very beginning, there was light. Down through time man has been **spiritually affected by real physical light.** Is there a reason for it?

How will the people of God appear in the future when they are with Him?

"And there shall be no night there; and they need no candle, neither light of the sun; for the Lord God **giveth them light** [illumines them]; and they shall reign forever and ever," (compare Rev. 21:23 where the city is lighted or illumined by God).

Revelation 22:5

"…the LORD shall be unto thee an everlasting **light**, and thy God thy **glory**… for the LORD shall be thine everlasting **light**.…"

Isaiah 60:19-20

"…the **glory** of the LORD shall be **revealed**, and all flesh shall see it together.…"

Isaiah 40:5

"For the earth shall be filled with the knowledge of the **glory** of the LORD, as the waters cover the sea."

Habakkuk 2:14

"And they that be wise shall **shine** as the brightness of the firmament [heavens]; and they that turn many to righteousness **as the stars** for ever and ever."

Daniel 12:3

"Beloved, now are we the sons of God, and it doth not yet appear what we shall be: but we know that, when he shall appear, we shall be like him; for we shall see him as he is."

1 John 3:2

"When Christ, who is our life, shall appear, then shall ye also appear with him in **glory**."

Colossians 3:4

"Giving thanks unto the Father, which hath made us meet [fit] to be partakers of the inheritance of the **saints in light**."

Colossians 1:12

There is no way for us to fully appreciate the quality of life Adam and Eve enjoyed before their sin. Neither is there any way for us to really grasp the excellent relationship they had with their loving Creator. We know they had a close fellowship with God in person. Their innocence may very well have been bathed in God's glory. Could that glorious light have been dimmed when sin broke their dynamic connection with Spiritual Life himself (like pulling the plug)?

"And they were both naked, the man and his wife, and were not ashamed." *Genesis 2:25*

"And the eyes of them both were opened, and they knew that they were naked.…" *Genesis 3:7*

If Adam's "light" was suddenly turned out, that may give some insight to a New Testament reality.

"But if we walk in the light, as he is in the light, we have fellowship one with another, and the blood of Jesus Christ his Son cleanseth us from all sin."

1 John 1:7

God's light is now revealed through Jesus Christ the Redeemer, and through His life-giving Word. To Adam at the beginning, the glory of redemption was not known, but he did experience a brilliant glory of God's manifest presence continually.

Think ! When we realize that the word "glory" signifies brilliant blinding light, does that magnify our sense of awe toward our Creator for what He did at the beginning and what He will do in the future?

From Glory to Glory

"But we all [believers], with open [unveiled] face beholding as in a glass [mirror] the glory of the Lord, are changed into the same image from glory to glory, even as by the Spirit of the Lord."

2 Corinthians 3:18

[Be careful not to confuse the meaning of "glory to glory" with Romans 1:17, speaking of spiritual growth "from faith to faith."]

Do you notice the parallels of what we have seen now ?

• God is clothed in radiant glory.

• The two men on earth who experienced the most transcendent manifestation of God's presence literally shone like lights because of the exposure to the heavenly glory.

• God's saints in the future will "shine like lights" because of their intimate presence with God.

THINK ! Is it likely that the original glory of Adam in the garden may have given him the appearance of being clothed in light? Could it be that their physical nakedness was insignificant compared to the brilliance of their spiritual likeness to God?

That phrase in 2nd Corinthians 3:18 says we are being transformed into the same image. What is that image? Was not man made in God's image? He lost that glory because of sin. He became "dead in sin." The glory that original man had was wonderful, but the glory to come, through our merciful Redeemer, Jesus Christ, is far more glorious!

The Destiny of God's Chosen People

The apostle Peter wrote:

"But ye are a chosen generation, a royal priesthood, a holy nation, a peculiar [special] people, that ye should shew forth the praises of Him who hath called you out of darkness into His marvelous light" (1 Peter 2:9).

When you don't know God, you are in spiritual darkness. If you don't take that first step of faith in the darkness toward Jesus, then you just keep groping in the confusion of hopeless darkness. Here is where the Living Word comes in and breaks up the deception of that seemingly endless darkness.

The Glorious Hope

The apostle Paul wrote in 2nd Corinthians 4:3:

"But if our gospel [good news of salvation] be hid, it is hid to them that are lost." The context is continuing from chapter 3, relating the experience of Moses covering his glowing face with a veil. The lost (described in verse 4) are those whose minds have been blinded by the god of this world (present age). Their disbelief actually prevents "the light of the glorious gospel of Christ" from shining upon them.

Notice how the analogy is intertwined with the Genesis account of creation in verse 6: "For God, who

174

commanded the **light** to shine out of darkness, hath shined in our hearts, to give the light of the knowledge of the **glory** of God in the **face** of Jesus Christ."

Where was the glory of God seen by the disciples on the Mount of Transfiguration? In the face of Jesus! Through whom can we be enlightened with the knowledge of God's glory? Again, it's Jesus.

Verse 7 concludes: "But we have this treasure in earthen vessels, that the excellency of the power may be of God, and not of us."

What is the treasure? The knowledge of God! Where is the treasure? In a very inglorious earthen vessel of flesh! That glory is there inside the believer in Jesus, but it has yet to be revealed and shine out. An Old Testament story tells us about other earthen vessels to beautifully illustrate the impact of this truth.

A Daring Adventure of Faith

The apostle Paul recognized a wonderful spiritual truth when he wrote the fourth verse of Romans chapter 15: "...whatsoever things were written aforetime were written for our learning, that we through patience and comfort of the scriptures might have hope."

When you just follow the life of a man like Gideon in the Bible Book of Judges, it is encouraging to anybody who feels like there's no hope. One particular event in his life shows a God-given symbolism worth noticing.

In Judges 7:16, we see Gideon leading his 300 brave and obedient soldiers against a seemingly invincible enemy army. Leave it to God to select the strangest weapons for their midnight surprise raid: trumpets in their right hands, and in their left hands they each held an old clay jug upside down over a burning torch. As they followed their leaders down from the hills into the valley of the enemy's camp, they blew the trumpets and smashed the earthenware vessels. God used the sudden surprise display to totally confuse the enemy and bring His people to a glorious victory. Perhaps you already see the remarkable parallel.

The Ultimate Victory

"We have this treasure in earthen vessels...." Perhaps you've noticed a certain brightness on the countenance of a consistent, joy-filled disciple of Jesus Christ. There seems to be a gleam of God's grace that literally brightens the room when they enter. For most of God's people down through time, the only way for the final glory of His presence to be revealed is through the death of this earthly "tabernacle." However, there is coming a day when God himself will lift the veil and translate His living saints from the vile realm of our earthly estate into the glorious presence of the King of kings and Lord of lords.

Our Assignment for the Present Time

In a figurative way, Jesus told those who follow Him: "You are the light of the world" (Matthew 5:13). Since the glory of God is veiled and unseen by those who are blinded in unbelief, we have been given a life-giving commission to change the situation (Matthew 5:16).

> "Let your light so shine before men, that they may see your good works, and glorify your Father which is in heaven."
>
> *Jesus, Matthew 5:16*

SECTION FOUR

Unlocking the Mysteries of Ancient Civilizations

"For enquire, I pray thee, of the former age, and prepare thyself to the search of their fathers..."

Job 8:8

Think of the vast span of historical events that have occurred on our planet since the beginning. Prepare yourself for a surprising twist you may not have heard from your history books.

Many people living today have very little concept of, and almost no sense of connection to the tremendous parade of people and cultures that have gone before us. Prepare yourself for a shocking revelation that goes cross grain to conventional thinking.

As the ancient record of Job encourages us, "get ready" for a discovery that will revolutionize the way you think about the family of mankind from a biblical perspective.

And remember…

…it was about 4,000 years ago that God himself declared that…

Nothing they plan to do will be impossible for them!"

What Is the Truth about Man's Primitive Past?

The 1828 *Webster's Dictionary* defines the word "primitive" as "pertaining to the beginning or origin; original; first as the primitive state of Adam; primitive innocence...." Modern dictionaries add "earliest or primary" but also add a second definition. See if it reflects an evolutionary bias: "resembling the manners or style of long ago: old-fashioned; simple; plain." Have we adopted evolutionary "baggage" in our concept of our earliest ancestors? Prepare yourself for a paradigm shift!

Why are ancient cultures mysterious?

There are many fascinating books and articles being circulated these days with titles that include the word "mystery." The marvelous findings of ancient civilizations are so surprising, they leave us puzzled without simple answers.

THINK! Have you ever stopped to think just why so many things turning up about ancient cultures end up being classed as "mysteries"?

What is the traditionally accepted concept of man's development on earth?

The popular evolutionary scheme regarding man assumes some unproven things to be true.

• Man supposedly began as an ignorant, brutish animal.

• The family of man gradually progressed through the "cave man" stage to become technologically modern, after tens of thousands of generations over millions of years.

• The most ancient beginnings of man's cultures are automatically thought to be "primitive" or crude in comparison to later developments.

• Objects produced by very early humans are expected to be "primitive" or simple.

• According to evolutionary notions, anything that is discovered to be both very ancient and technologically or culturally advanced, is automatically labeled as a mystery.

Think! Is it true that:
Ancient = Primitive or Simple???

Some Facts to Keep in Mind

1. As we learned in our discussion on "missing links," there is **no evidence to even suggest, let alone prove, that man has evolved from any form of beast.**

   The missing links of evolution are still MISSING!

2. As we learned in section two, there is **no evidence to prove the earth is millions or billions of years old!**

   • Such an idea is a fabrication of evolutionary speculation.

   • Many strong evidences point to a very young age of the earth, in terms of a few thousand years and not millions.

   The earth is very young!

Evolutionary View of Man's Achievements

According to evolutionary thinking, man began at his lowest and gradually evolved upward in technological skills, physiological abilities, wisdom, and culture. Of course this supposedly happened over a vast span of millions of years that no one has even the slightest clue about.

The Biblical View of Original Man

What does the Bible say about ancient man? Psalm 8:5 declares clearly that man was made by God just a little lower than angels. The original word translated there as "angels" is the Hebrew word "ELOHIM." It is used throughout the Old Testament to refer to Almighty God himself.

According to the Bible, man began at the top and then fell into depravity because of rebellion (sin) against his Creator.

The reality of all verified history, including our most trustworthy history book ever (the Bible), depicts man at his best in the beginning. He has been on a downhill slide ever since.

Passage of Time

The evolutionary view:

A slow, and gradual rise in technology, wisdom and culture over millions of years.

The creationary view:

A rapid drop from perfection just 6,000 years ago with periodic revivals in technology.

What does all verified history tell us about early man?

Keep in mind that our goal is to discover the facts! We want to get information that is verified. We are interested in theories only insofar as they can be backed up by the facts.

The well-known fact of man's existence on earth is that his history goes back only about 5,000 years. The ancient Hebrew Scriptures are not the only record of this. Other early cultures testify to a similar starting point in time. Anything proposed to go back before 5,000 years ago is sheer speculation based on the insistence of evolution and the rejection of true history.

"He that answereth a matter before he heareth it, it is folly and shame unto him."

Proverbs 18:13

Is modern man really more advanced than ancient man?

Unlocking the Mysteries Of Creation

What Are Typical Patterns of Ancient Cultures?

Down through history man has repeatedly risen to high peaks of civilization. These great epochs of man's past are acknowledged in the Bible. Abraham's home city of Ur in Mesopotamia was a sophisticated cultural center. Egypt in the time of Moses was a magnificent high point in human achievement. Babylon in the days of Nebuchadnezzar and the prophet Daniel was one of the proudest and most accomplished societies that ever lived on earth.

The Bible mentions a number of other ancient cultures whose splendor has only recently been revealed by the archeologist's spade. The Hittites, the Phoenicians, the Syrians and the Persians are a few of the well-known cultures that have come to light in more modern times.

Of course there were many other ancient civilizations on earth that are not mentioned in the Bible. The Bible was never intended to give a complete history of the world.

What characterized ancient high cultures?

Besides the fact that advanced ancient civilizations achieved tremendous material accomplishments, there is another quality that consistently typifies each one. Their rulers led what biblically qualifies as an "anti-Christ" system of tyranny.

When you examine the remains of these cultures you find evidence of widespread violence and gross immorality. At the apex of these societies, demonic manifestations and human sacrifice were commonplace.

Above: Life was cheap and slavery was common in Mesopotamian cultures like Persia, where these winged bulls guarded the gate to their city.

Right: In the Yucatan, the Mayans sacrificed thousands to their pagan idols on altars like this.

What has been the end of every ancient civilization?

Virtually every ancient society on this planet reached a point of destruction and ruin. Despite their great accomplishments, their submission to spiritual wickedness in high places ultimately brought their downfall.

The Bible records how God had to bring catastrophic judgments to put an end to the degradation and violence. Other cultures around the world have also been destroyed, sometimes by mysterious catastrophe.

Even though depraved cultures have been wiped out, man himself has always managed to rebuild again somewhere. Knowing man's capacity and drive to dominate, it's easy to realize why cultural revival has always recurred down through the centuries.

Knowledge Will Increase in the Last Days

Israel's notable prophet Daniel told (in chapter 12 verse 4) that knowledge would increase in the last days. He predicted this almost 3,000 years ago, at the height of one of the world's greatest cultures. The Babylonian Empire was majestic beyond our comprehension. Early descriptions of the great city leave us in awe at the accomplishments of that high culture. Mankind has certainly had some low points since then. The dark ages of Europe lasted hundreds of years. But man has certainly risen again to great heights of accomplishment in the last few generations.

What about cave men?

When you think of primitive simple tribes, you usually think of groups that are found in remote places today. Are they your ancestors? Many of them are descendents of advanced races like the Mayans and Egyptians. Likewise there were backward tribes in the past; offshoots of main line human culture.

Stone age people are living today while others explore space and invent computers. Society has its dropouts today. So it was in the past too, while others advanced culturally and technologically. Diversity of lifestyles and sophistication are common to every era of history.

Does the Bible mention cave men?

The prophet Isaiah observed an interesting thing 2,600 years ago. Notice Isaiah 2:19:

> *And they shall go into the holes of the rocks, and into the caves of the earth, for fear of the Lord, and for the glory of his majesty, when he ariseth to shake terribly the earth.*

In the face of devastating judgment, survivors find shelter wherever possible. Caves would be logical habitations for many years. What most people today fail to realize is that there have been several widespread catastrophes in the ancient past that have literally reduced civilization to rubble. Because this reality is so obscured in modern culture, a deeper historical study is in order. Seek out research findings on catastrophic history and ancient cultural remembrances of these events.

The Ishtar gate of Babylon

What Were the Capacities of Original Man?

The Bible declares plainly that God made man in His image – in God's class. Man was intended to be the dominator – a peaceful majestic king over all the creation. To better appreciate the implications of such a position, and the potential of man in his beginning, we need to ask a thought-provoking question.

THINK! When God created the angels, how much time do you suppose it took before they were fully functional?

It's no problem at all for us to recognize that the angels of God operated on full potential at once, from the beginning of their existence. But what do we think about man? Could it be that our concepts of early man have been severely warped because we are not thinking through the implications of God's Word?

Did the earliest humans have to grope around in ignorance for hundreds or thousands of years before they finally woke up enough to figure out how to make a wheel or build a fire?

Just how smart was original man?

Many of us have heard that a mature human uses less than 10% of his brain capacity in his lifetime. You no doubt know some who get by on a whole lot less. The brain has the uncanny ability to store the equivalent of millions of bits of information every second while it automatically regulates many of the delicate chemical functions of the entire body.

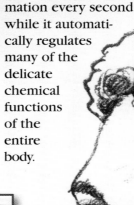

In terms of memory capacity, each of us has the ability to see, hear and feel a new "snapshot" of experience every second for a million years and record every detail with precision. No computer can match the human brain. All that information is stored but some of us have a harder time than others trying to retrieve it. Yet all it takes is a whiff of a certain aroma to bring back a whole flood of memories of people and events from our past.

THINK! If evolution were true, why would we have a huge percentage of brain power – perhaps 90% – that we will never use in our entire lifetime? (Remember that evolution presumes traits arise for usefulness.)

Explanations of origins that leave God out of the picture have no clue how a severely handicapped child can occasionally become what is called a "savant." Some have incredible ability to calculate long strings of mathematical functions instantly without thinking about it. Others with no formal training can play a complex piano concerto after hearing it once, and remember it for years. Could these be a hint of the innate abilities placed in man at the beginning by our Creator?

An Intellectual Exercise for Original Man

An event at the dawn of human history provides some insight to the mental ability of the first man. The Creator gave Adam the commission to name all the animals. Don't you think they were meaningful names? How would we do with such a job? Judging by the way we name our pets with words like spot, muffin and fuzzy, you get the feeling that modern imagination has suffered some severe setbacks.

Some think that Adam might have spoken an early form of Hebrew. Whatever language it was, do you think the names he gave the animals would have been every bit as significant and meaningful as the names we use in our own languages? The nobility of the lion, the delicacy of the flamingo, the size of the elephant, and the cunning of the fox were no doubt reflected by the names Adam gave. We know they were acceptable names because God accepted them without question. In Genesis 2:19, we are told: "And whatsoever Adam called every living creature, that was the name thereof." God saw no reason to quibble with Adam about his choice of words.

Is Technology a Modern Innovation?

When Adam fell into sin, his world fell with him because he was given control of it. But even in his fallen state the family of man accomplished things in ancient times that were far from "primitive."

What defines technology?

When we think of advanced technology we automatically load it down with all sorts of complicated hardware... costly things that only few people know how to tinker with.

From the very beginning, why wouldn't ancient man have always been:

• Inventive...

• Technical...

• Creative...

• Productive...

• Ingenious?

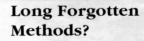

The Science of Agriculture

In Genesis chapter 4 we see that Abel, first son of Adam and Eve, was a shepherd. His brother Cain was a farmer. A descendant of Cain is described in verse 20 as a keeper of livestock.

If you think about it, animal husbandry and agriculture are quite advanced forms of human enterprise. They require technology in specialized tools and processes, as well as a lot of planning and care. If you think raising cattle or sheep is a primitive occupation, go out and try it for a season. The business is far beyond some mythical "stone age."

Long Forgotten Methods?

In *Science Yearbook 1980*, produced by World Book encyclopedia, an article is titled "Farming The Negev Desert." This arid wilderness in southern Israel is springing to life with lush crops like onions and peaches. Why?

As the lead caption pointed out: "Israeli scientists have revived long forgotten farming methods to grow crops by storing the sparse desert rain in the once barren soil." Was this some complex technological invention?

All they did was build little dikes around sloped plant basins to collect the sparse rain. Brilliant? Indeed! Expensive? No way! It was an ancient method, but very effective. How would our "modern" style try to grow crops in a desert?

To begin, we would build huge dams hundreds of miles away. Pumping stations and pipelines costing millions of shekels would be built to move the water. Hydroelectric generators would be built to run the pumping network. Since dry years are anticipated, oil burning power plants must provide a back-up source of energy, assuming they were smart enough to not let all the water out of the dams.

The more complex our modern technology gets, the more vulnerable we, as a society become to things like: natural disasters, wars, economic turmoil, environmental pollution, and terrorism. Our "high tech" world has made us dangerously dependent on specialized experts and a steady supply of exotic resources to keep everything running smoothly.

Could the ancients have had better ways of doing things?

Consider the impact of capitalizing on natural laws to a greater degree than we do. Are there ways to produce larger long-term benefits with less complex hardware? We tend to think that our 21st century technology is the best way to go. But as we explore the clues to ancient people we find they had a different, and quite possibly wiser approach. For lack of an established definition on this area of study, let's consider the implications of a new term:

Simplified Ingenuity!

Here's a thousand-year-old ruin of the Mayans in the tropical Mexican city of Palenque. When I was there on a sweltering April day the temperature was about 100 degrees and the humidity was 100 percent. But inside this immense six-story-high structure, the rooms were air-conditioned! Carefully engineered corridors channeled cool air through massive stone passageways, assuring ideal temperatures year round. That is **Simplified Ingenuity!** How would we air-condition a building in the tropics? Would our methods be... technical? costly? short-lived? This ancient system has been working perfectly for a thousand years, even with no one there to maintain it!

There is nothing new under the sun!

How Early in Man's History Did He Build Cities?

In Genesis 4:17, we find that Cain found his wife, fathered a son, and built a city. Before we can explore the implications of building a city so early in history, we must examine another issue here. Because Cain was the eldest son of Adam and Eve, the first humans on earth, a question is often asked of those who believe the Bible…

Where did Cain get his wife?

Most of us assume from reading Genesis that Cain and Abel grew up together as the only children of the first parents. Skeptics insist there had to be other families on earth besides Adam's. Where else could Cain get his wife? Does the Bible offer a reasonable explanation?

THINK! What is the evolutionary alternative? Where did the evolutionary "Cain" find his sweetie pie? Follow their reasoning! The origin of male and female anything has no naturalistic explanation. It's another case of one of those evolutionistic fairy tales: "There just happened to be a female that evolved from some creature with different chromosomes, and she just happened to appear at the same place in time to meet her 'man.'" Now look out… we're going to have to talk about the birds and the bees here for a minute.

Let's hear the rest of the story!

Take a look at Genesis 5:4. Note Adam fathered "other sons and daughters." That means, along with the three that are mentioned, they had at least seven kids. Now, honestly, how many kids do you think Eve could have produced? Remember, these people were perfectly made by God. And what did He command them to do? Multiply and fill the earth… as with children? Yes! God is no killjoy. He invented all the beauty and pleasure of sexuality and marital love. It's sin that has warped the "eros" part of humanity: lust, greed, pride, and lack of self-control. But God gave many sexual aspects to humans that He did not give to animals. Ask a knowledgeable doctor or contact a good creation resource center. The human body is awesome and God deserves our deepest sense of awe for creating it.

So you live in a paradise, you've got the most beautiful creature on the planet for a mate. You've got all the sensory equipment to enjoy doing the job right…. And if there ever was a command that humans were able to easily obey, it had to be this one!

Hmm… Nice place, but I'm gettin' kinda lonely…

Don't you suppose Eve might have delivered a baby every other year for a decade or two? What a woman!

Large families used to be common. After all, there's security in numbers… more hands to work the garden and tend the flocks. How many kids were in your great grandma's family? So, what did God give Eve after her son Abel was killed? Another son, Seth! And how old was Adam when Seth was born? 130 years old! Now, how many kids and grand kids and great grand kids, could Adam and Eve have in 130 years? Lots! So, was finding a wife a problem for Cain? Note also that all these folks lived for hundreds of years in those days.

The genetic pool was obviously its purest at that time. Genetic deterioration took centuries to take its toll. Marrying a sister or half sister was not genetically dangerous like it is now. Notice that Abram married his half sister Sarai. It wasn't until Moses' time that God forbade marrying close relatives.

Cities before the Flood?

So Cain built a city! Notice it doesn't say a mud hut! This first son of the first man built the first city on earth – the city of Enoch. Imagine that. And even then you'd expect a "city" to have houses and shops, and streets and lanes, and some kind of government. Yes, knowing people, politics must have gotten an early start.

Some people might have a problem with the idea of cities existing even before the Flood, but think about it. The very earliest relics of human history soon after the Flood indicate the existence of cities. You might expect that. After all, settlers in the new world tended to rebuild something like the culture they left behind in the old world.

So what have archeologists dug up?

If you do a search through some of the magazine articles published in the last few decades about ancient cities, look what you find. The popular *Readers Digest* magazine published a string of articles with these titles:

- "Palenque: Mexico's Mysterious Lost City"
- "In the Beginning: The Mystery of Ancient Egypt"
- "Mysteries of the Maya"
- "Unsolved Mysteries of the Great Pyramid"

THINK! Why are these references to very advanced ancient cultures that produced cities labeled "mysteries?"

The area of the world known as the "Middle East" includes what has often been called the "cradle of civilization." Egypt, Israel, Syria, Iran, and Iraq are just some of the modern nations of that region. Early empires there built cities like Jericho and Babel over 4,000 years ago! The remains of the most ancient cities testify to some amazingly advanced accomplishments despite their great antiquity, including indoor plumbing and piped hot water.

What Are Some Genesis Examples of Early Technology?

Primitive Music?

Early mankind, before the Flood, knew the technical skills involved in the production and use of musical instruments. That should come as a surprise to those of us who have been taught that the technology of music was invented at a much later time.

In Genesis 4:21, we read that a pre-Flood descendent of Cain "was Jubal…the father of all such as handle the harp and organ [or flute]." Modern scholars realize that some of the most ancient musical harps come from the oldest culture on earth. The workmanship of Sumerian craftsmen produced not only exquisite artistic designs, but also technologically precise instruments. Some are over 3,500 years old.

Gilded Sumerian harp decoration from prior to 1500 B.C.

Traditional Ideas?

What is the usual picture painted for us about the evolutionary discovery of something like music? It might go something like this…

Grandpa Ramapithecus was out with the family picking berries one day. As he reached for a juicy looking berry dangling overhead, he accidentally stumbled over a vine of some sort that was stretched rather tightly between two trees. He just happened to notice the curious twanging sound he hadn't heard before. He turned back to pluck the vine again. Twa-a-a-ang… it did it again. He grunted excitedly to his fellow berry-pickers nearby who had gotten sidetracked picking termites from the ground and lice out of each other's hair. Just about when he was going to announce his discovery of music, he realized he hadn't yet invented language!

Sure, this is just a spoof, but for an evolutionist, it's not too far off the party line of their pipe dreams of how advancements had to happen accidentally.

Music in Ancient China

A Chinese zither, similar to our modern autoharp, was surprisingly found recently in a tomb dating back over 21 centuries. Also in the tomb was a mouth organ with 22 carefully designed bamboo pipes, a mouthpiece and fingering stops to operate it. The details were finely crafted and in no way could be called crude or lacking ingenuity.[1]

According to the Chinese people there was an emperor living 4,600 years ago who wanted to standardize musical sounds. He sent his servant into the mountains in search of a special bamboo pipe that produced a middle C note when cut to a specific length. From that pipe all the other notes of the musical scale were mathematically derived.[2]

What about metallurgy?

Genesis 4:22 also mentions specifically that another descendent of Cain was "Tubal-Cain, an instructor of every artificer [craftsman] in brass and iron."

The Iron Age is ordinarily thought to have begun about 1000 B.C. The so-called Bronze Age started another 2000 years before that. But this reference claims that both bronze and iron were manufactured way before that, even before the Great Flood.

From the Danube River basin in Europe have come sophisticated copper tools dating back at least 4,000 years. *National Geographic* magazine included a surprising article in the November 1977 issue titled: "Ancient Europe is Older Than We Thought." The newest findings are changing ideas about the "barbarians." The author wrote: "Scholars once thought that metallurgy spread here from the Near East, but [there is] … proof [now] … that a metal industry was fully developed in Europe when it was just starting in the Aegean [i.e. Greece]."

Ancient Metal Works All Over the World?

The oldest large-scale metallurgical factory in the world was unearthed in 1968 at Medzamor, in Armenia. Over 200 furnaces reveal that an unknown society of 4500 years ago produced ornaments and weapons of copper, lead, zinc, iron, gold, tin, manganese, and 14 kinds of bronze. Several pairs of tweezers from over 3,000 years ago are made of such exceptionally high-grade steel, that scientists marvel at this industrial age near Mount Ararat.[3]   Since Ararat is the landing place of Noah's ark, would you expect that some of the earliest technology might turn up there?

In ancient Egypt discoveries of elaborate goldsmith work date back more than 4,000 years. Some metal artifacts surviving from ancient Egypt appear to have been electroplated! That is, gold has been applied over base metals through a process of electrolysis.[4]

From a 22-century-old emperor's tomb in China comes this report in *National Geographic* for April 1978: "Metal swords [buried for 2200 years, have been] treated with a preservative that prevented corrosion for [all that time]" The ancient weapons were alloyed of tin, copper, and 13 other elements including magnesium, nickel, and cobalt.

Why Search Out the Former Age?

The book of Job is one of the most ancient preserved pieces of literature on earth! Apparently first written long before Moses assembled the early records of Genesis, Job lived somewhere in Arabia before Abraham arrived in Canaan. It was very different then, 4,000 years ago. Rivers, pastures, and even ice are reported in a region that has now degenerated to a barren desert. But before that, how had things changed even more? Look what Job recommends.

> **"For inquire, I pray thee, of the former age, and prepare thyself to the search of their fathers."**
>
> *Job 8:8*

So what is the purpose in this exhortation to examine the citizens of an even more ancient time?

> **"For we are but of yesterday, and know nothing, because our days upon earth are a shadow."**
>
> *Job 8:9*

The author of this ancient book implies people long before Job's time had far greater knowledge than Job's culture.

THINK! What does he mean by telling us that we are ignorant because our "days on earth are a shadow?"

What is the "former age?"

If Job lived more than 4,000 years ago, and the Flood destroyed all civilization just a few hundred years before that, could the "former age" be the era before that great dividing point of history?

The Flood radically altered earth's environment. The atmosphere and cosmic protection were seriously damaged. A significant result of that was the severe reduction in the lengths of human lives. Compare the lengths of the lives of the men living before and after the Flood.

How long did the earliest humans live?

Men Born Before the Flood

Adam	lived 930 years
Seth	lived 912 years
Enosh	lived 905 years
Kenan	lived 910 years
Mahalalel	lived 895 years
Jared	lived 962 years
Enoch	lived 365 years*
Methuselah	lived 969 years
Lamech	lived 777 years
Noah	lived 950 years**

Enoch (*) did not even die because he walked so closely to God that God finally just took him. See Genesis 5:24 and Hebrews 11:5.

Noah (**) lived 600 of his 950 years before the Flood. The Bible makes no apology for the ages of these patriarchs. There is no way to equate these years to anything but ordinary solar years for the text to have continuity.

What could people do in 900 years?

THINK! What could be accomplished if you lived the equivalent of 13 life spans of 70 years?

With that kind of time on your hands you could devote a hundred years to building a nice business and then take off a hundred years to perfect your favorite musical instrument skills. Think of the trips you could plan. You don't suppose you'd just sit around all that time would you? Imagine the gardens you could design or the projects you could build!

With a growing population all speaking the same language, it wouldn't be very many generations before there would be millions of people for you to interact with.

Mankind's Greatest Handicap?

In the 4,000 years since the Flood, man's decreasing longevity has been his most serious handicap. Just when you grow up enough to start figuring out how to live right you start realizing how brief your stay on the planet is. Yet look at what some people have accomplished in such short life spans. Think of an inventor like Thomas Edison. How about the genius of a Leonardo da Vinci? Consider the fertile creativity of a Bach or Beethoven. Now multiply that potential and you'll start to grasp the impact of what Job is urging you to do.

Men Who Lived After the Flood

Shem	lived	600 years*
Arpachshad	lived	438 years
Shelah	lived	433 years
Eber	lived	464 years
Peleg	lived	239 years**
Reu	lived	239 years
Serug	lived	230 years
Nahor	lived	148 years
Terah	lived	205 years
Abraham	lived	175 years

Shem (*) was only 100 years old when the Flood came.

Peleg (**) was born at the time of another worldwide catastrophe (the division of the continents) and his life span began a trend that dropped further descendents to about half that of his post-Flood ancestors. Something changed again.

The Compound Value of Sharing Life with Multiple Generations

On the following page you will notice how the lives of pre-Flood men overlapped for hundreds of years. Noah lived 600 years with his grandfather. Do you think they just sat around and played dominoes all that time? But the story gets more exciting when you realize Noah's grandpa, Methuselah, likely knew Adam for over 200 years! Think of it: sharing the insights and discoveries of the first man—the one who walked with the Creator in the Garden of Eden and talked with God face to face.

People who live long know much!

"With the ancient is wisdom: and in length of days understanding."

Job 12:12

With all the natural resources available to them why wouldn't man before the Flood have exercised his natural human ingenuity every bit as much as he has in the last four thousand years? Developments in commerce, industry, the arts, politics, science and medicine would have all been possible, and even inevitable, before the Flood.

Unfortunately, how often have you seen typical religious art depicting Noah using quaint old tools to build the largest wooden ship in history? Do you see how evolutionary brainwashing has warped our concepts of ancient man?

Baloney detector

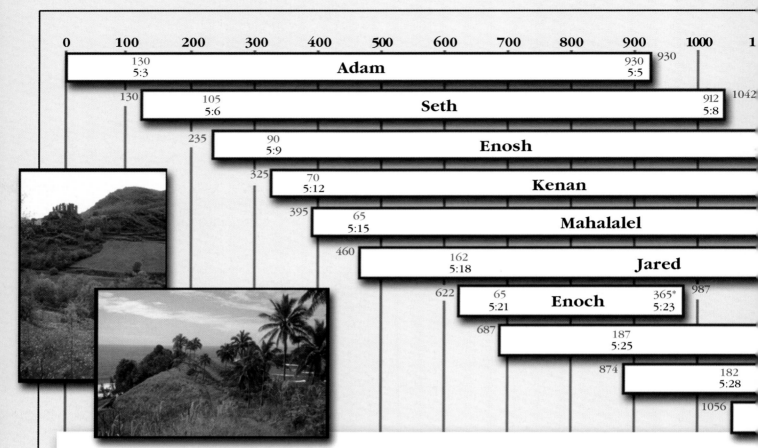

Chart values (timeline by year A.C., top scale: 0, 100, 200, 300, 400, 500, 600, 700, 800, 900, 1000):

- **Adam** — 130 / 5:3 ... 930 / 5:5 ... 930
- **Seth** — 130 | 105 / 5:6 ... 912 / 5:8 ... 1042
- **Enosh** — 235 | 90 / 5:9
- **Kenan** — 325 | 70 / 5:12
- **Mahalalel** — 395 | 65 / 5:15
- **Jared** — 460 | 162 / 5:18
- **Enoch** — 622 | 65 / 5:21 ... 365* / 5:23 ... 987
- 687 | 187 / 5:25
- 874 | 182 / 5:28
- 1056

Imagine the Accumulation of Ancient Wisdom over the Generations

The chronological life spans of the Genesis patriarchs reveal many surprising facts.

How to use the chart:

- The vertical lines mark off centuries starting with Adam's creation in the year zero.

- The bar for each individual represents his total length of life compared to others in the chart.

- The first number (green), at the left corner of each bar is the year after creation (A.C.) of the man's birth (Enosh: born 235 A.C.)

- Inside each bar the first number from the left (light blue) shows the age at which he fathered the son beneath him; the reference below that is where that information is found in the Book of Genesis.

- At the end of each bar the top number (red) is the age of the individual at death. Beneath that is the scripture reference where that information is found.

- The number at the end of each bar (purple) is the year (A.C.) of death.

- The Bible records of these men's lives use specific numbers and words to establish a clear continuous family record from Adam to Jacob.

- The age of each man is given when the next named son (or possibly grandson) is born. Thus there can be no gaps in the time line even if direct descendents are left out of the list.

- From Adam to Noah's father, the record tells that other sons and daughters were born. Thus there were at least five children per family. Statistically, the population of earth by the time of the Flood (1656 A.C.) could easily have numbered three billion people.

- Some references indicate that the name "Methuselah" means "when he dies—it shall be sent [judgment]." Look to see why this is significant.

* Enoch did not die.

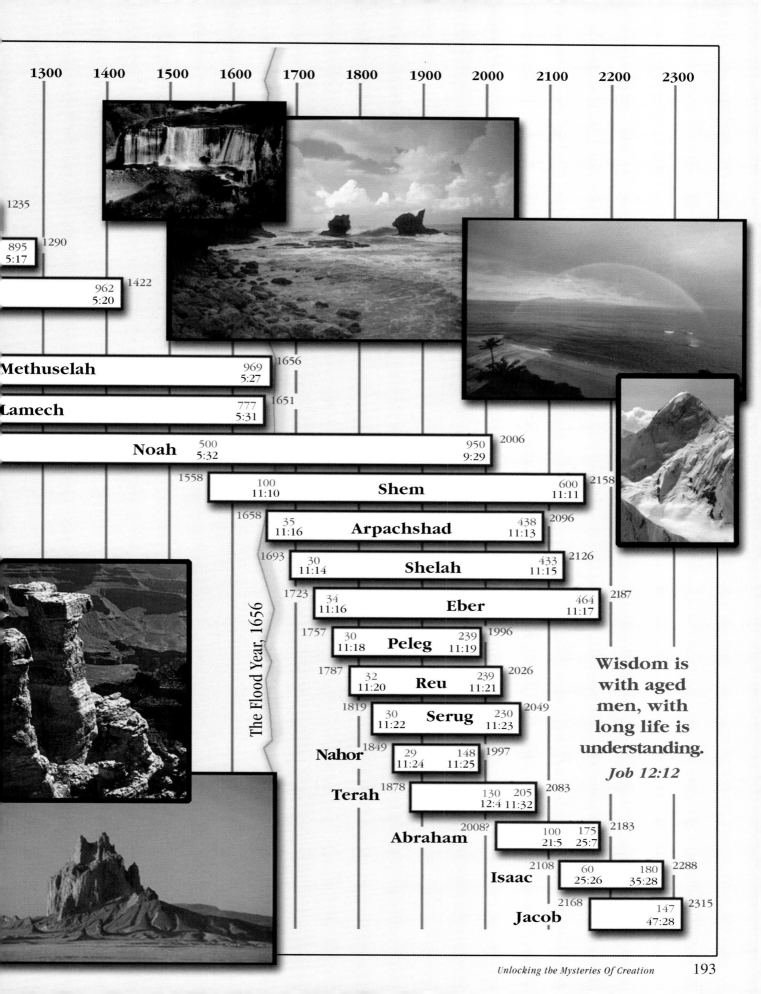

1300   1400   1500   1600   1700   1800   1900   2000   2100   2200   2300

1235

895
5:17
1290

962
5:20
1422

Methuselah   969 / 5:27   1656

Lamech   777 / 5:31   1651

Noah   500 / 5:32   950 / 9:29   2006

1558   100 / 11:10   **Shem**   600 / 11:11   2158

1658   35 / 11:16   **Arpachshad**   438 / 11:13   2096

1693   30 / 11:14   **Shelah**   433 / 11:15   2126

1723   34 / 11:16   **Eber**   464 / 11:17   2187

1757   30 / 11:18   **Peleg**   239 / 11:19   1996

1787   32 / 11:20   **Reu**   239 / 11:21   2026

1819   30 / 11:22   **Serug**   230 / 11:23   2049

Nahor   1849   29 / 11:24   148 / 11:25   1997

Terah   1878   130 / 12:4   205 / 11:32   2083

Abraham   2008?   100 / 21:5   175 / 25:7   2183

Isaac   2108   60 / 25:26   180 / 35:28   2288

Jacob   2168   147 / 47:28   2315

The Flood Year, 1656

Wisdom is with aged men, with long life is understanding.

Job 12:12

What About the Spiritual Insight of the Ancients?

In the early centuries after the Great Flood (which happened about 2400 B.C.), the ancient influential resident of Arabia named Job had amazingly advanced insights. His book shows that he had considerable knowledge of agriculture, building trades, and the laws of nature. He even had remarkable knowledge of spiritual truth too.

"Oh, that my words were now written! Oh, that they were printed in a book!"
Job 19:23

Skeptics programmed by evolutionary thinking used to publicly scorn the idea that even Moses could write intelligently in 1500 B.C. Here we see a man living another 500 years earlier and he's contemplating printing a book! But he feels so strongly about his message that he even wants to chisel it into granite.

"That they [my words] were graven with an iron pen and lead in the rock for ever!"
Job 19:24

Jeremiah 17:1 says, "The sin of Judah is inscribed with an iron pen with the point of a diamond." Is it possible that 20th century industry is not the first to have diamond-tipped iron-cutting tools? Obviously Job wants to make a very indelible record and share this bit of knowledge with others for a long time to come. So what is Job's important message?

"I know that my redeemer liveth, and that he shall stand at the latter day upon the earth: and though after my skin worms destroy this body, yet in my flesh shall I see God."
Job 19:25-26

THINK! He is talking about the return of the Messiah and the physical resurrection of the redeemed. How could some ancient tribal chieftain understand such a deep theological concept even before Moses wrote the Torah?

One of the feeble arguments among Bible critics asserts that men invented the concept of God and the supernatural only after he evolved enough to produce such an elaborate scheme. The very idea of "redemption" is a complex and advanced element of systematic theology. The plan of a coming Redeemer was familiar to Job because he had a continuous link of ancestors back to the man who received it from God himself in the Garden. But don't forget that God is able to inspire men with unnatural knowledge through His Holy Spirit.

"The secret of the Lord is with them that fear him: and he will shew them his covenant."
Psalm 25:14

Job, a man who lived 4,000 years ago, had insights that stun modern anti-biblical philosophers. Why? Could it be that they are rejecting history as well as the privilege of the personal revelation of God's gift?

Where did Job get these lofty ideas?

Look at the chart on pages 192-193. It becomes clear that families could easily pass down and reinforce important information for many generations. Noah's family carried a tremendous link of traditional knowledge from the old pre-Flood world to the new age in which Job lived. Noah's grandfather, Methuselah, no doubt told him about his own father Enoch. He left this world nearly 70 years before Noah was born.

In the New Testament Book of Jude (verse 14 & 15), we read that Enoch, who lived with Adam, predicted the coming of the Lord "with ten thousands of his saints to execute judgment upon all, and to convince all that are ungodly among them of all their ungodly deeds...."
Enoch lived before the Flood. He probably had greater prophetic insight than all our modern prophecy teachers and mystics.

What did original man have?

By now you have no doubt begun to add some new dimensions to your understanding of the original man and his vast family. The Bible does put things in a fresh perspective doesn't it? As Psalm 119:130 says:

"The entrance of Thy Words giveth light."

Isn't it amazing to realize that our greatest grandfather, Adam, actually shared intimate fellowship with the Creator? **Original man was so totally different from the secular slant we are commonly programmed to believe!** Imagine what his superior sharpness could do with the benefit of conversing openly with God.

Adam and Eve had...

- Personal harmony with nature
- Open communication with God
- Peaceful dominion over the animals
- The ideal physical bodies
- Intimate awareness of natural laws
- Intimate response to spiritual laws

God's masterpiece of creation, man, was made to be the master of the entire earth. He didn't have to be some phony kind of "superman" in physical strength to do it either. With his God-like spirit, Adam and his family had the power to accomplish anything to which they applied their ingenious minds and corporate energy. Anything? Note that even after the fall into sin, the Torah records that God said of mankind:

"Nothing that they may propose to do will be out of their reach."

Genesis 11:6

God himself made this statement years after the Great Flood, at the time of the confusion of the languages at the Tower of Babel. If men can conceive of a way to accomplish a goal, they will be able to reach their goal! Why should we think that only in modern time has mankind reached the level of scientific wisdom where this is really possible?

Such a sweeping assertion seems almost too impossible, even if God did say it. But keep in mind that God made man "just a little lower than Elohim [God himself]." So let's look for the evidence to support this biblically realistic approach to ancient human achievements.

The Heavens Declare
the Glory of God!

*"When I consider thy heavens, the work
of thy fingers, the moon and the stars,
which thou hast ordained...."*

Psalm 8:3

Do you know someone raised in a city who finally went camping in the mountains to witness a moonless midnight sky for the first time? The initial shock of stellar majesty inspires a new appreciation for the word "awesome." Truly, that amazing sight is over-powering to our human emotions and intellect. But is there more to it than that?

> **"The heavens declare the glory of God; and the firmament sheweth his handiwork. Day unto day uttereth speech, and night unto night sheweth knowledge. There is no speech nor language, where their voice is not heard. Their line is gone out through all the earth, and their words to the end of the world."**
>
> *Psalm 19:1-4a*

The most ancient cultures on earth were captivated by the stars. To them the magnificent annual procession of stars had historic and prophetic significance as well as usefulness in scheduling their agricultural seasons. Today most of us have essentially zero knowledge of what those ancient people knew of the stars. For generations, the "speech" the Psalmist reported 3,000 years ago, and the "line" that reached the "end of the world" has been missing from our education.

THINK! Though corrupted cultures from Nimrod and Babel forward have wrongfully worshiped the stars, associated them with mythical legends, and developed systems of pagan astrology, we must ask a basic question:

What was the Creator's purpose for decorating the night sky with gleaming gems?

> **"And God said, Let there be lights in the firmament of the heaven...and let them be for signs, and for seasons, and for days and years."**
>
> *Genesis 1:14*

"Days and years" clearly reveal God's intent that stars would help us measure time. Indeed, the precision and regularity of stellar observations from the beginning have provided man with annual "skymarks" in time. Like the microscopic world, the starry heaven shows us God's proclivity for order and design that allows us to rely on scientific measurement. The creation is not random, chaotic and unpredictable, otherwise modern scientific investigation could never have developed.

What is a sign?

Just as the Hebrew word used here suggests, we understand that a sign is a pointer or a mark to indicate something other than itself. The same word is used when Moses' staff changed into a snake as a sign to Pharaoh that God's ambassador was speaking God's message to him. The Hebrew word here for seasons is something fixed or appointed, such as the time God set for Sarah to have her son, but never used to suggest climatic seasons like summer.

THINK! The stars were not scattered by some massive explosion. Psalm 8:3 indicates God ordained them.... He placed them where they are for a purpose!

Is the purpose of the stars revealed by their names?

Did you know that God named the stars? Psalm 147:4 says: *"He telleth the number of the stars; he calleth them all by their names."* God assigned Adam to name the animals, but He reserved the naming of the stars for himself. Do you suppose star names might have profound meanings?

God's Personal Nature Study

Four thousand years ago, God asked Job many questions including some about specific stars and well known star groups.

> **"Canst thou bind the sweet influences of Pleiades, or loose the bands of Orion? Canst thou bring forth Mazzaroth [the 12 signs called constellations] in his season? Or canst thou guide Arcturus with his sons?"**
>
> *Job 38:31-32*

Orion is universally known as a heavenly sign. God authenticates its unique name in the Bible. Orion means "Coming as light." It is one of several star groups that identify a mighty man coming to triumph over the great enemy of mankind – the serpent!

What is the first prophecy in history to promise the Redeemer for mankind, alienated from God by sin?

Cursing the serpent in the Garden of Eden after Adam was found trying to hide his newly revealed naked- ness, God prophesied in a graphic word picture:

> **"And I will put enmity between thee [the serpent embodying the Devil] and the woman, and between thy seed and her seed; it shall bruise thy head, and thou shalt bruise his heel."**
>
> *Genesis 3:15*

Before Job, before Babel, and before the Great Flood, fathers recounted the heavenly portrayal of Orion to their sons. The coming Prince of Light holds a great club in his right hand and the token of his victory in his left – the head and skin of the "roaring lion" that is Satan. The sign shows Orion's left foot is raised to crush the head of the enemy.

If there is any doubt about the identity of this Orion, all we need is to examine the ancient names of the notable stars in the figure. The brightest (in the right shoulder of Orion) is named Betelgeuz, which means "the coming of the branch." Next is Rigel, seen in the foot poised over the head of the enemy. It is the seventh brightest star in the sky. Rigel means "the foot that crushes." In the left shoulder of Orion is

Bellatrix, which means "quickly coming" or "swiftly destroying." One of the three stars in Orion's belt is named Al Nitak, "the wounded one," and the star in his right leg is called Saiph, meaning "bruised" (the very word in Genesis 3:15).

THINK! Should it surprise us to learn that Orion is the most brilliant of all the constellations?

If the "glory of God" was intended to be declared by the stellar heavens (Psalm 19), is it likely that their story line, in the 12 signs called Mazzaroth by God himself in Job 38, signifies the eternal plan of Messiah, who alone is "the brightness of His glory"? (Hebrews 1:3).

The Hebrews, the Aztecs, the Babylonians, the Egyptians, the Assyrians, and the Chinese all have the same 12 constellations with the same essential meanings. To look at the stars, there is no reason to group them as they are in the zodiac… **except if it was divinely designed to tell a story.** Modern astrology is simply a **perversion** of God's original intention. And the real meaning of the zodiac is truly an awe-inspiring revelation.[5]

The Real Meaning of the Zodiac

God himself, in Job 38, draws attention to "*Mazzaroth*" (Hebrew word for the zodiac). How significant is this? The names and sequence of the 12 signs plus their 36 decans or supporting symbols, seem too precise to be any less than God's invention. Since God really named the stars, and providentially kept those names in the traditions of cultures worldwide, it makes sense why the Devil has worked hard to distract people from the truth of God's primary purpose in creating the stars – for signs. Yet when we analyze the evidence, the truth shines through the pagan distortions. For a student of the Bible the graphic symbols in the sky are unmistakable.

An ancient tradition recorded by Jewish historian Josephus says Adam's son Seth and great-great grandson Enoch were first to tell the prophetic story in the stars. As the pageant unfolds, two major characters are clearly seen in a variety of "signs." One is the adversary, "that old serpent, the Devil." The other is the glorified "seed of the woman" and Redeemer of mankind. Three prominent constellations present the serpent with names like Draco, Serpens, and Hydra. The enemy is often shown attempting to kill the Redeemer or thwart His mission. Yet we see the biblical Christ glorified in the stars as the innocent sacrificial Lamb, the Provider of endless, life-giving water to His own, and the conquering Lion. It's truly an awesome panorama.

The Beginning and Ending of the Gospel

When you begin with the constellation Virgo, the Virgin, and end the stellar circle with Leo the lion, the panorama really comes into focus. The bright star in Virgo's left hand is Spica. Its ancient Hebrew meaning is *The Branch*, but unlike 19 other Hebrew words for branch, this word alone is reserved in scripture for the Messiah. Jeremiah (23:5,6) calls the Branch the *King*. Zechariah reveals Him as the *Servant* (in 3:8) and as *Man* (in 6:12), while Isaiah identifies the Branch only with *The Lord* (4:2). Likewise the gospels reveal the Messiah as King (in Matthew), as Servant (in Mark), as Man (in Luke), and as God (in John).

Associated with Virgo are her three decans: Coma, Centaurus, and Bootes. These famous constellations depict the *Desired Son*, the *Despised Sin-Offering*, and the *Coming Shepherd*. Following these are ten meaningful groups of constellations before we come to the last one named Leo the Lion. The tail of the lion touches the head of Virgo. And what is that kingly Lion doing? He is victoriously pouncing on the head of the serpent Hydra, whose body stretches across a third of the heavens! As Revelation 5:5 depicts Christ as the Lion of the tribe of Judah coming in glory to defeat "that old serpent, the Devil," so we see the precision of the Creator's intent that these stars would be for "signs." The Hebrew word for Leo refers to a lion hunting down its prey. In the powerful front legs of Leo is the first-magnitude star **Regulus** meaning *Treading Underfoot*. The next brightest star is **Denebola** in the tail and means the *Judge Who Comes*.

Unlocking an Ancient Symbolic Mystery

Exploring these symbols in the sky we begin to unlock another long-hidden mystery, hidden that is to secular scholars who are ever learning but resistant to the knowledge of the truth. The strange mystery of the ancient Sphinx is revealed at the 4,000-year-old zodiac of Dendereh, found in the ceiling of the portico of the temple of Esneh in Egypt. There, between the signs of Virgo and Leo is a picture of the Sphinx, with the head of a woman and the body of a lion. When we understand the divine story of redemption that begins with "the seed of the woman" (Genesis 3:15) and ends with the triumphant lion (Revelation 5:5), the Sphinx makes sense.

A casual consideration of the names and symbols of the 48 constellations shows that the story of the cosmic conflict between the Serpent and the Deliverer is the main theme throughout the entire narrative. It appears again and again in different ways and can hardly be the result of naturalistic accident or human creativity. The most ancient zodiacs are likely closest to the original. Important differences are seen then. For example, Gemini may originally have been a united man and woman, rather than two male twins. Also, Cancer, the Crab, seems originally to have been an enclosure into which people came from all sides.

A fascinating and revelatory journey awaits anyone willing to study the historical data gathered by excellent authors on the subject of Mazzaroth. Request *The Gospel in the Stars* (1884) by Joseph Seiss, and other books, currently available. The complementary revelations referred to in Psalm 19 seem to verify both the special story line of the stars and the testimony of the Lord's written Word. Since the antediluvian world saw the stars and knew their names, and since we see that revelation dimly but have the written Word, we can see the Messiah portrayed in the sequence of constellations in their natural order.[6]

The greatest story ever told is magnificently foretold by the Creator's awesome design of stellar signs. The Enemy has perverted that story and blinded many to its life-changing power, but those willing to dig out the hidden truth will be royally rewarded with a personal revelation that transforms them forever.

1. Virgo – A deliverer supernaturally conceived of a virgin, born as a man, yet the Son of God.

2. Libra – To balance the scales of divine justice, an adequate price must redeem man from his curse as a sinner.

3. Scorpio – Redemption's price requires the death of the deliverer, who must also destroy the serpent-tempter.

4. Sagittarius – The two-natured Deliverer conquers the Dragon and receives praise.

5. Capricornus – The living fish results from the death of the atoning sacrificial goat, and resurrection life is shared with all true believers.

6. Aquarius – The Water of life poured out on the redeemed who await the sure return of the Redeemer.

7. Pisces – Delayed salvation sets free those in bondage who await their coming King.

8. Aries – The Lamb was slain to deliver the freed woman to marry her Deliverer, overcoming the great dragon.

9. Taurus – The Judge comes to rule as Prince of Glory as fiery wrath pours out on His enemies and His people are protected in the day of wrath.

10. Gemini – The coming Prince of Glory crushes the enemy underfoot.

11. Cancer – The gathered flock is secure in the safe harbor and rest of their Coming Redeemer.

12. Leo – The triumphant Redeemer destroys the Serpent upon which the wrath of God is poured out and then is devoured by the raven.

What's the Solution to the Puzzle of Advanced Ancient Technology?

As modern man has probed around his world, he has encountered thousands of fascinating discoveries showing evidence of high degrees of culture and technology. In our time, these puzzling relics reveal that ancient people were advanced far beyond what evolutionary dogma expected.

Some Examples of Ancient Advancement

- The Great Pyramid at Giza in Egypt
- The accurately placed megaliths (large cut stones) of Stonehenge, England
- Tremendous ancient cities like Babylon and Teotihuacan in Mexico
- Knowledge of mathematics and astronomy, as found in the Mayan cities of the Yucatan
- Even perplexing hints of technical tools, machinery, electricity, and air power

When evolutionists insist that primal man was stupid, how do they explain all this?

Have you ever heard of the book *Chariots of the Gods* by Erik Van Daniken? Popularized in the 1960s and 70s, the "alien invasion" theory has deceived millions of people. Their idea is that since earthlings must surely have been too "primitive" to do all this on their own, the obvious explanation is that superior beings brought the technology here from distant worlds in outer space. The absurdity of this premise becomes very clear in light of scripture and known astronomical observations.

"They did not like to retain God in their knowledge."
Romans 1:28

"Because they received not the love of the truth, that they might be saved. And for this cause God shall send them strong delusion, that they should believe a lie: that they all might be damned who believed not the truth, but had pleasure in unrighteousness."
2 Thessalonians 2:10-12

What is the Bible's perspective on ancient technological achievements?

To find insight, let's explore the wisdom of the wisest man who ever lived (besides the Messiah). A thousand years before Christ, he was the king of Israel for 40 years during the highest golden age of the nation. First Kings 3:12 describes Solomon as supernaturally gifted by God with a wise and understanding heart. There was none like him ever before or ever after that time. He was king but he was also a naturalist and writer who studied all manner of things. He wrote an essay that became the book of Ecclesiastes in the Bible. His introduction gives us a key to understanding the real world.

Nothing New Under the Sun

"The thing that hath been, it is that which shall be; and that which is done is that which shall be done: and there is no new thing under the sun.

"Is there any thing whereof it may be said, 'See, this is new?' It hath been already of old time, which was before us.

"There is no remembrance of former things; neither shall there be any remembrance of things that are to come with those that shall come after."

Ecclesiastes 1:9-11

THINK! What is Solomon saying to us from a vantage point of 3,000 years ago, 1,000 years after Abraham and 3,000 years after God created Adam? His perspective is the very center of all human history!

Solomon insists that there is nothing new under the sun and neither will future generations come up with anything really NEW! Discoveries we think are 'new' may have been known long ago. Do you think he's referring here to the citizens of that "former age" that Job mentioned?

The wise king observes that people are very forgetful. Maybe that's because we don't live long enough anymore, or more likely because we are brainwashed by an ignorant revision of the past.

THINK! Name some common things or skills of the past that are almost forgotten by much of society today! Craftsmanship, music, logic and manners are just a few areas to get you started.

Can you think of some new invention that soon became obsolete because of another invention? Just when we really think we've arrived with some state-of-the-art discovery, we soon fulfill Solomon's prediction. There's "no remembrance of things to come by those who will come after." Could we be simply repeating the discoveries and inventions of another age, a forgotten society that could even have surpassed our own?

Remember! God is the one who declared that:

Nothing they plan to do will be impossible for them.

If you fail to learn from the past you are in danger of repeating it.

Why E.T. Can't Be...

THINK! Since life is impossible on all our neighboring planets, where is ET's home?

• Life requires many precise physical conditions. It's a miracle that life exists even on earth.

• The nearest star is 4.5 light years from earth. At a million miles an hour you'd have to travel 3,000 years to get here. And some have difficulty believing Adam lived 930 years! Only problem is Alpha Centauri is just a big hot stellar mass... no life could exist there.

• There is no clear proof that any planet orbits some distant star. The search is urged by evolutionists who insist planetary systems and life can evolve again and again, even when the evidence defies that evolution ever happened at all.

• Most UFOs are still unidentified. The phenomenon captivates many who'd rather contrive vain imaginations than heed God's call to **prove all things.**

• Many UFO reports can be explained by advanced human technology and some natural phenomena such as ball lightning.

• If reports of contact with "aliens" are taken seriously, just examine the results: fear, superstition, bizarre tampering with animals, associations with witchcraft, and a cultic preoccupation with weirdness.

• Reports of some UFOs claim motions that defy natural laws: changing direction instantly while traveling at supersonic speed or mysteriously disappearing.

• To defy natural laws could qualify UFOs as "supernatural." There are two types: evil demons or godly angels.

• When gullible people say UFOs are angels, simply ask: "Do they direct humans to worship the Creator?" True heavenly angels can do nothing less!

• What conclusion does that leave you? (Request more information on this topic from CRF.)

When Did Civilization Begin?

The Great Deluge is the greatest turning point preserved in the historic traditions of almost every culture on earth.[7] Noah and his sons made sure that all their descendents knew the story well. When they finally scattered from the Tower of Babel they took the memory of the Flood with them to remote regions like North America, where Flood legends still survive today. The most ancient societies of the Middle East arose in the first century after the Flood (by 2300 B.C.). They had no primitive ancestors! The advanced civilizations they built were vain attempts to regain the lost world of the former age.

Yet what did these survivors and their descendents create?

Wonders of the Ancient World

The Great Pyramid of Cheops

By the modern (though comparatively primitive) city of Cairo, Egypt, the most magnificent structure on earth baffles any attempts to explain the brilliance of those who built it. In sheer mass, the Great Pyramid is the largest single edifice ever built. This lone survivor of the Seven Manmade Wonders of the World is also the most perfect building on the planet! Its intricacy and stability defy explanation. After 4,000 years the foundation hasn't even sagged! The planning and construction evade our attempts to understand it. Over two million quarried blocks weigh an average 2.5 tons each. An unlikely crew of 100,000 men could not have set them, even if a new block was positioned every two minutes for 20 years![8] It was already centuries old when Abraham arrived in Egypt from his city of Ur.

THINK! With no evidence of a crude past, why were the earliest people of Egypt so culturally sophisticated, yet insistent that their forefathers were even more advanced?

In May 1975, *Readers Digest* published the article: *"When did civilization begin?"* It noted, *"The new findings have made a shambles of the traditional* [edit. evolutionist] *theory of prehistory."*

The intellectual Greeks were aware of earlier civilizations destroyed by natural causes. Philo of Alexandria (c. 30 B.C. – 40 A.D.) wrote, *"By reason of the constant and repeated destructions by water and fire, the later generations did not receive from the former the memory of the order and sequence of events."* Plato recorded in the *Timaeus* that his ancestor, Solon, wrote: *"There have been and there will be again many destructions of mankind,"* and when civilization is destroyed, *"you have to begin all over again as children."*[9]

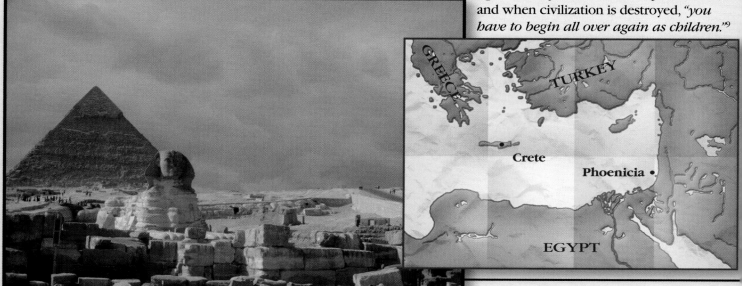

The Minoan Civilization

The Mediterranean island of Crete was home to a quite modern civilization until an immense volcanic catastrophe wiped it all out 3,500 years ago. Archeological discoveries of the culture show that the Minoans were advanced in:

- government
- the arts
- social structure
- language
- medicine
- mathematics
- astronomy
- sea travel

Their destroyed city had brightly colored columns, polished marble floors, and intricate architectural decoration, much like our classic government buildings today. Their delicate pottery rivals the finest of any culture since.

Discoveries that Startled the World

The *National Geographic* magazine (2/78) reports the amazing finds of Crete's early archeologists, Heinrich Schlieman in 1873 and Sir Arthur Evans in 1900. They were the first to reveal the surprises of Cretan culture dating back to 1650 B.C. The impressions from their discoveries reflected an obvious preconceived stereotype about ancient man. They are reported as saying that this culture was: *"a remarkably sophisticated society for so distant a time."*

THINK! Why would sophistication seem "remarkable" in a context of antiquity? Would such a comment be made by a person with the insight and knowledge of men like Solomon or Job?

The Sumerians founded the first civilization we know of after the Flood. They recognized a pre-Flood civilization predating their own. The "Weld Prism" discovered in 1922 and now in the Ashmolean Museum at Oxford contains a history of a scribe from 2100 B.C. After recording the ten pre-Flood kings, his writing ends with the poignant words, *"and the Flood overthrew the land."* Later the Babylonians and Assyrians also recognized the pre-Flood era as a source of superior literature. One Babylonian king recorded that he *"loved to read the writings of the age before the Flood."* Assurbanipal, who founded the great library of Nineveh, also referred to the great *"inscriptions of the time before the Flood."*[10]

Why did the pre-Flood and post-Flood civilizations advance to such impressive accomplishments so rapidly? They seem to have made much better use of their intellect than most of us do today. Most of us can't even remember important information without the help of gadgets and written records. Also consider that with shorter life spans after the Flood, the compiling effect of shared knowledge would be diminished. Facts were more easily distorted. Written records were more needed than before, to retain information, and compensate for the loss of superior mental ability.

In the last century or so, many fragments of obscure technological artifacts of unknown origin have surfaced. Though they may have been reported in local papers or even scientific journals, they have been largely ignored by modern science as oddities or anomalies because of the mystery they represent.

Do Artifacts of Very Ancient and Even Pre-Flood Technology Exist?

Since the Bible's account of history is truly confirmed by studies of world cultures and the earth sciences, then is it so hard to accept the implications of an object that indicates that very ancient humans were technologically and culturally advanced? We need not be puzzled by, or cynical about, the reported discoveries of "Ooparts!"

Out Of Place Artifacts = o. o. p. arts

Our modern minds are filled with baggage of distorted teaching about the past. We tend to be baffled by the accidental discoveries of ancient artifacts found deep in the geological layers of the earth, yet there are many of them. Ooparts have always been found by ordinary people who were not looking for them nor thinking about careful scientific documentation. Unfortunately, many intellectuals tend to disqualify them for "proofs" of anything. So they are called "anomalies." But why not let them stand on their own merit?

"Blown from the rock...there is no deception in the case"

The June 1851 issue of *Scientific American* (Vol.7, p. 298) reported a metallic vase that was dynamited from solid rock in Dorchester, Massachusetts. The story, printed first in the *Boston Transcript*, said:

"…the two parts together …formed a bell-shaped vessel, 4 1/2 inches high, 6 1/2 inches at the base, 2 1/2 inches at the top, and about an eighth of an inch in thickness. The body of this vessel resembles zinc in color, or a composition metal, in which there is a considerable portion of silver. On the sides there are six figures of a flower, or bouquet, beautifully inlaid with pure silver, and around the lower part of the vessel a vine, or wreath, inlaid also with silver. The chasing, carving, and inlaying are exquisitely done by the art of some cunning workman… There is no doubt but that this curiosity was blown out of the rock…. The matter is worthy of investigation, as there is no deception in the case."

The editor of *Scientific American* stated that the object must have been made by Tubal-cain, the Genesis founder of metallurgy. After circulating from museum to museum, it was eventually lost or hidden from public view.[11]

Mysteries to Evolution… but Inevitable from Genesis

The discoveries of many puzzling artifacts have been reported in local papers and scientific journals over the past few centuries. Some are recorded in recent reference books like *Ancient Man: A Handbook of Puzzling Artifacts*, by William Corliss. Here are just a few.

A gold chain was found embedded in a chunk of coal when Mrs. S.W. Culp of Morrisonville, Illinois was stoking her kitchen stove. The local paper reported it in the June 9, 1891 edition, saying, "This is a study for the students of archaeology who love to puzzle their brains out over the geological construction of the earth from whose ancient depth the curious are always dropping out." By the way, the coal that is best explained by a global flood in historical times is typically described by evolutionists as being around a hundred million years old. So it is not allowed to contain objects of human manufacture.

A gold thread was discovered in 1844 in a quarry in England, embedded in granite judged by geologists to be 60 million years old. The London *Times* sent investigators who reported their opinion that the thread had indeed been of artificial manufacture.

In 1885, foundry workers in Austria broke a block of coal and found inside a small metal cube that was examined meticulously at the Salzburg museum. The 2 1/2-inch cube was alloyed from steel and nickel and weighed 1.73 pounds. The edges were perfectly straight and sharp; four sides were flat, but two opposing sides were convex. A deep groove had been cut all around the middle of the cube. There was no doubt that the cube was machine-made and seemed to be part of a larger mechanism.

The Nampa Doll

In the 1880s at Nampa Idaho, a discovery was made creating quite a stir at that time. While drilling a water well through 300 feet of soil, clay and lava rock, a small pumice-stone statue came up from the hole. It was a carefully carved image of a human form. Examining the tiny particles of sand cemented to the image, experts reported then that it could not have been recently fabricated and that it could not be a hoax. The scientific report in 1889 continues: "The high degree of art displayed in the image is note-worthy. It is not the work of a boy or of a novice. The proportions are perfect, and there is a pose of the body that is remarkable, and which differentiates it from anything that has been found among the relics of the Mound Builders… It supports… that civilization advanced on the Pacific Coast long in advance of that which has anywhere else been discovered. And it is by no means impossible that we have some relics of those catastrophes by floods which are so universal in the traditions of all nations."[12]

The London Artifact

In June of 1936, Mr. and Mrs. Max Hahn of London, Texas saw a portion of a broken wooden handle poking from the gravelly sandstone ledge by the Red Creek near the Llano River. They dislodged a hefty chunk of the sand-stone that firmly cemented the object. In 1946-47 their son George broke open the chunk, exposing 60 percent of the metal hammerhead of the tool that was still firmly embedded in the stone. The wood of the handle near the iron head is partly turned to coal, but otherwise is fibrous old wood with some crystals filling the center where it is broken. Oddly, the iron hammerhead was not rusted but smooth with a brownish

fossil coating, apparently formed by the surrounding rock. A half-inch groove was later made on the metal with a steel file. It showed the inside metal to be bright silver in color and well hardened. The tool appears to be that of an artisan. The six-inch-long head has a cross pattern, tooth-like surface on one end, and a cut post on the other that looks like it had a soft material of some kind wrapped around it.

The rock formation that encased the hammer is supposed by evolutionary reckoning to be 135 million years old. The concretionary sandstone formation extends over a wide area with no evidence of being re-mixed or filled in at a recent date. A search was done to find a similar counterpart tool made in recent centuries. None was found. Battelle Labs in Columbus, Ohio did an analysis of the metal. Strangely, the report observed 96.6% iron, 2.6% chlorine, and 0.74% sulfur, but oddly no carbon which is normally needed to harden iron. Metallurgists discussed this anomaly and indicated that an alloy of iron with chlorine cannot be made in earth's present atmospheric condition. A tomographic analysis of the hammer was done in 1992 at Texas Utilities. It showed the metal of the hammer was superior quality with no inclusions or irregularities…clearly the result of high-tech metallurgy, unavailable until the space age. Because the idea of a Golden Age before the Great Flood is deliberately ignored (just as Peter predicted) some modern geniuses have figured the logical explanation is that some alien from space landed here 135 million years ago and left his hammer behind.[13]

THINK! If geologists are wrong about the dating of rock layers and historians are wrong about the advancement of original man, finding a sophisticated artifact in coal or sandstone should help us unlock the mystery with a biblical perspective on the Deluge and our rediscovery of antediluvian technology!

When Was the Golden Age?

When you think about the great accomplishments of human civilization down through time, consider the question: "When was the Golden Age for all the various major cultural centers on earth?"

When was the Golden Age of Greece?

When in Greek history did the culture experience its most classic high point? It certainly isn't today is it? But if you go all the way back to the exquisite culture of Athens around 500 B.C. you find the greatest period of culture Greece has ever known.

Italy had a Golden Age once too.

From 200 B.C. to A.D. 200 the Roman Empire achieved world-class proportions as one of the greatest cultures on earth.

What about Mesopotamia?

We're speaking about Iran and Iraq – the lands of the Tigris and Euphrates rivers – which are a mess culturally today. It's been a cultural wasteland for over 2,000 years, but go back to the days of Daniel and Esther, the Persians and Babylonians, and you'll find a rich heritage before 500 B.C.

When was the Golden Age of Central or South America?

Again, Mexico and the smaller nations continue in poverty and political unrest in modern times. But between 1,000 B.C. and A.D. 500. the Mayan, Olmec and other cultures experienced tremendous accomplishment, not to mention the Aztecs and Incas of later times.

Egypt's Golden Age ended with the Exodus around 1500 B.C. All the dynasties of the great pharaohs were even earlier.

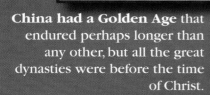

China had a Golden Age that endured perhaps longer than any other, but all the great dynasties were before the time of Christ.

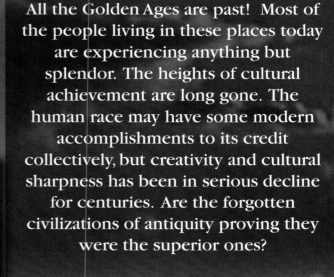

All the Golden Ages are past! Most of the people living in these places today are experiencing anything but splendor. The heights of cultural achievement are long gone. The human race may have some modern accomplishments to its credit collectively, but creativity and cultural sharpness has been in serious decline for centuries. Are the forgotten civilizations of antiquity proving they were the superior ones?

China's Forgotten Connection to Genesis

Who did the Chinese people worship during the nearly 2,000 years before Buddhism, Confucianism and Taoism showed up about 500 B.C.? The oldest continuing culture on earth has preserved in its traditions and language one of the most remarkable confirmations of the Genesis record of history.

Abraham's El Shaddai and China's Shang Ti

Like all others, the Chinese people and language originated at the tower of Babel. Since the Chinese language is symbolized by idea-graphs or pictographs, parts of many words have significance revealing an origin that cannot be confused. The pictograph for "tower" contains symbols leaving little guesswork about its origin. Remember the language originated **after** the event at the Tower of Babel.

Mankind + One + Mouth = **United** + Grass = **Undertake** + Clay = **Tower**

Mankind, being of one speech (language) united, despite the fact they were under the curse and all would die like grass, undertook their clay brick project – the tower!

The Chinese themselves built no towers or pagodas until the Buddhist era. This amplifies their long respect for the reality of the event at Babel as the source of their word for tower.[14]

The ancient Hebrews, descending from Noah and Abraham 2,000 years B.C., learned that one of the names of the Creator is El Shaddai, their "provider." The ancient Chinese worshipped a single Creator God named Shang Ti, the "emperor of heaven." Similar sounds? Same God?

The Ancient Annual Ceremony that Mystified Confucius

For over 4,000 years, Chinese emperors traveled every year to the border of the country or the border of the capital city to make a sacrifice – the Border Sacrifice. This annual ceremony began before the first Chinese emperors over 2,000 years before Christ, and continued until A.D. 1911 when revolutionaries overthrew the Manchu emperor. The main act of this ceremony was the burning of the entire body of a young bull on an altar in the open air as a burnt offering to God. Puzzled by the meaning of the border sacrifice, Confucius (551-479 B.C.) wrote: "He who understands the ceremonies of the sacrifices to Heaven and Earth...would find the government of a kingdom as easy as to look into his palm."[15]

Part of the emperor's recitation during the ceremony resembled the language of Genesis:

Of old in the beginning, there was the great chaos, without form and dark. The five elements [planets] had not yet begun to revolve, nor the sun and the moon to shine. In the midst thereof there existed neither form nor sound. Thou, O spiritual Sovereign, camest forth in Thy presidency, and first didst divide the grosser parts from the purer. Thou madest heaven; Thou madest earth; Thou madest man. All things with their reproducing power got their being.... Thou didst produce, O Spirit, the sun and the moon and the five planets....[16]

Does this sound like paganism or a reverence for the One True Creator?

THINK! Where did Adam make his offerings to God and teach his sons to do the same? It was likely under the shadow of the Cherubim, guarding the way to the tree of life at the east gate of the Garden of Eden.

The Chinese pictograph for "border" is a garden with four rivers radiating from its center. Beside it are three graceful lines representing Shang Ti, the Creator God. The bent line at the right is a person bowing before God. An older meaning for this character is "to come before God."[17]

From the Shang dynasty (1766-1123 B.C.) to the Ch'ing dynasty (1644-1911 A.D.), the Chinese traditionally regarded their history and writing with reverence, believing that these ancestral truths came from a divine source.[18] The wealth of information researched by medical missionary Dr. Ethel Nelson with Chinese scholars is well worth searching out in her books, *The Discovery of Genesis* and *God's Promise to the Chinese*.[19]

A simple example connecting the Chinese language to Genesis is the word for boat. The left radical means "ship" or "vessel." The upper right one is the number "eight." The lower right one is "mouth" or "person." What is the first boat mentioned in the Bible? How many "mouths" were aboard? Genesis 7:13 tells us: *"In the selfsame day entered Noah, and Shem, and Ham, and Japheth, the sons of Noah, and Noah's wife, and the three wives of his sons with them, into the ark."* So Noah's ark was a vessel with eight people.

Genesis 1:2 says the Spirit of God "brooded" or "hovered" over the face of the waters on the earth prior to the creation of land. The top of the Chinese idea-graph for "Spirit" shows the radical for "water" combined with "cover," giving the idea that God's Spirit "covered" or "hovered over" the waters.

This combination becomes the word "rain." In the Bible, rain is often used as a picture of how God pours out His Spirit upon the earth. Joel 2:23,28 says, *"He will cause to come down for you the rain, the former rain, and the latter rain …And it shall come to pass afterward, that I will pour out My spirit upon all flesh."*

In the middle of the "Spirit" graph, we see the radical for "mouth" ("person") repeated three times – reflecting the Trinity. Lastly, we see a radical known as a worker of magic. It shows three persons working together under heaven upon the earth.

The word "covet" or "desire" pictures a woman making a decision between two trees. Genesis 2:8-9 shows us that there were two significant trees side by side in the garden from which to choose. Genesis 3:6 reveals the woman made the wrong choice.

Two trees are at the top of the glyph for "forbidden." The bottom means "God commands or notifies." So the word "forbidden" means "God makes a commandment about two trees." In Genesis, God first forbade them to eat of the "tree of the knowledge of good and evil" (Genesis 2:16-17). After they fell to temptation, He forbade them to eat of the "tree of life" (Genesis 3:22-24).

The Chinese language predates the Hebrew of Moses' time by at least 700 years, yet it contains an accurate record of events in Genesis. Isn't God awesome to preserve such a record for us?

舟	Vessel
八	Eight
口	People/ Mouths
船	**Boat**
一	Heaven
冖	Covering
氺	Water
雨	**Rain**
口口口	3 Persons
巫	Miracle Worker
靈	**Spirit**
林	Two Trees
女	Woman
婪	**Desire**
林	Two Trees
示	God Commands
禁	**Forbid**

4 Rivers, 1 Garden   God   Worshipper

Border

Speak   Dust   Life   Walk

告 土 丿 辶

To Create

造

Who Really Discovered the New World?

For many years the modern Western mindset held the opinion that Columbus discovered America in 1492. Later it was conceded that Nordic Vikings from Scandinavia reached North American shores about 1000 A.D. We've been programmed to think that only modern discoverers traveled the world in relatively recent times, but was ancient man really so confined?

Did ancient man traverse the oceans?

THINK! If ancient civilizations were as technically capable as we are discovering here, then why shouldn't we expect to find evidence that they traveled all over the globe?

The March 1973 issue of *Readers Digest* magazine included an article titled: "Who Really Discovered the New World?" The article surprised readers with the following statement:

"Startling evidence reveals that the American continent drew many early visitors, including some more than 2,000 years before Columbus!"

Maps of the Ancient Sea Kings?

In 1929, a cleaning crew discovered an old torn map in the palace of a Turkish sultan. It turned out to be a surviving document from a lost collection of ancient maps compiled in 1513 by the Turkish admiral and pirate named Piri Reis. The carefully hand drawn map on gazelle skin referenced the now lost maps of Columbus and other maps that date back all the way to the time of Christ. The information on this map has stirred a controversy of historic proportions.

20th century cartographic experts have pointed out that the Piri Reis map shows accurate details of the coastlines of Africa and South America that should not have been charted until the invention of the navigator's chronometer in the 18th century. Even more surprising are the details showing part of the western coastline of Antarctica with no ice on it, plus the land bridge between South America and Antarctica that is now under water.

The famous Piri Reis map of 1513 was drawn with the benefit of ancient maps that predated Columbus, yet it shows geographic details of the coastlines of not only Africa (on the right) but also South America and Antarctica (on the left).

Harvard-trained historian and cartographer, Charles Hapgood, wrote his important book, *Maps of the Ancient Sea Kings: Evidence of Advanced Civilization in the Ice Age*, in 1966. It's no wonder his conclusions about the Piri Reis map are controversial. He wrote that these old maps were, *"the first hard evidence that advanced peoples preceded all the peoples now known to history."* He concluded, *"that some ancient people explored the coasts of Antarctica when its coasts were free of ice."* [20]

Antarctica shows prominently on the map made by Oronteus Finaeus in 1531. The details of the south polar continent are surprisingly accurate, even showing mountain ranges and rivers that are now under ice more than a mile thick. Some were not discovered until 1957 using sonic reflection devices to "see" beneath the glaciers.

Nothing New Under the Sun?

Was Antarctica free of ice at some ancient time? … a time when people were capable not only of traversing the oceans, but also of careful charting of geographic features? This goes contrary to evolutionary dogma insisting Antarctica has been ice-bound for over a million years. The technology needed to do all this must be very advanced, but it seems to have been forgotten. Isn't that exactly what Solomon predicted?

Who visited America before Christ?

If you're up to reading a highly controversial perspective, find the book, *America B.C.,* by Harvard University professor Barry Fell. Doctor Fell graphically documents archeological finds showing there were many global travelers before the time of Christ. Locations are shown all along the Atlantic coast of America and Canada, and even places flanking major rivers of the interior. Artifacts and huge stone monuments called dolmens indicate that even Celtic people came to America as well as Iberians, Egyptians, and Libyans.

One small example of the kind of thing analyzed by Dr. Fell is an inscribed stone found in West Virginia in 1838.[21] The engraved writing on the stone couldn't be deciphered when it was discovered, but now it has been identified as a dialect of the Phoenician language used in Spain before Christ.

THINK! How did an ancient Old World artifact arrive in North America long before the 15th century explorers? Is it satisfactory to simply disregard these discoveries because they don't fit the evolutionary assumptions about ancient man?

A 90-ton dolmen supported by five peg-stones at North Salem, New York. It is linked to a Celtic-Iberian king because of a similar monument near Dublin, Ireland.

Ancient Sea Travel

Did ancient civilizations travel to diverse parts of the globe? The traditional evolutionary dogma insists that such accomplishments had to wait for modern technology in relatively recent centuries. But what does the evidence show us?

The Ancient Polynesians

The ancient Polynesian people are routinely thought of as a rather primitive society. Yet they are famous for their vast travels all over the enormous Pacific Ocean.

By the time of Christ, these aggressive explorers had colonized every habitable island in the Pacific Ocean. They had methodically covered over 15 million square miles of open sea.[22] These mariners had impressive observational skills. They knew how to navigate by sighting the comparative rising positions of stars. They understood the currents of the oceans and the barely perceptible ocean swells. Both of these phenomena have been discovered by western civilization only in what we call "modern times."

Mysterious Easter Island

One place the Polynesians colonized is what we call Easter Island. It is well known for its megalithic heads carved out by a stone age society. These great monuments have been called mysteries by modern scientists. Some of the massive heads, smoothly sculpted in iron-hard volcanic rock, weigh over 180,000 pounds (90 tons). Most of them are 12 to 15 feet in height. One is 32 feet tall.[23] How did they get here?

Transported several miles from the quarry to their platforms, these stone age masterpieces have been a puzzle to modern researchers. How were they carved and moved? Their immensity is a challenge even to modern engineers and equipment. What was the purpose of these strange figures, gazing soberly across windswept, treeless plains? The society that made them is gone forever, but what they left behind makes us wonder if they knew much more than we usually credit to them.

2,300 miles off the coast of Chile, remote and barren Easter Island is populated by at least 1,000 of these strange giant heads.

The World of Mexico Before Columbus

There is more evidence of early world travel that comes from ancient Mayan centers of Mexico and Central America.

Numerous ceramic and stone artifacts show clear characteristics of African and Oriental people. They seem to have had contact with the ancient citizens of Mexico. Even Phoenician features show up on an ancient ceramic cup from Guatemala.

Hard Heads in the Face of Hard Evidence

It's amazing that evidence like this means absolutely nothing to those "experts" who stubbornly deny the possibility that ancients could travel worldwide. Many of these educated aristocrats insist that the only way humans ever could have arrived in the Americas is by slow migration across the Bering land bridge from Asia to Alaska during the last ice age. They say these nomads slowly wandered southward over generations before they finally settled in Mexico and began the development of one of the most technologically advanced cultures the earth has known: the Mayans.

What Insight Does the Bible Offer on Ancient Global Sea Travel?

In the Old Testament book of First Kings (10:22), we read about King Solomon's shipping enterprise. This great king joined forces with neighboring King Hiram of Phoenicia. Together they sent merchant ships on voyages lasting three years. They brought back all sorts of exotic imports from distant lands.

The Phoenician neighbors of King Solomon have been called the "Sea Lords of Antiquity." They were masters of the ocean. We know they traveled all over the Mediterranean world and even down the west coast of Africa. The trip across to the east coast of South America is less than the distance from Israel to the Rock of Gibraltar at the entrance to the Mediterranean Sea.

THINK! Why couldn't the "Sea Lords of Antiquity" cross over to South America? Only one reason –

The evolutionists won't let them!

The Atlantic Ocean is a mere lake when compared to the Pacific Ocean. If the Polynesian people mastered the Pacific over 2,000 years ago, why couldn't the mighty Phoenicians manage to cross the tiny Atlantic? Isn't it amazing how prejudice can blind intelligent people to something so obvious?

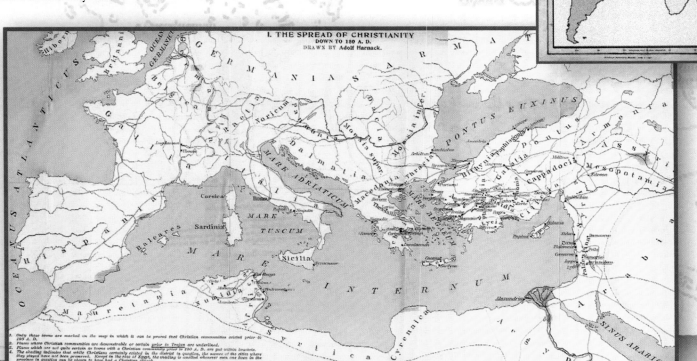

I. THE SPREAD OF CHRISTIANITY
DOWN TO 180 A.D.
DRAWN BY Adolf Harnack.

Why Are Ancient Pyramids a Worldwide Phenomena?

"And they said, 'Go to, let us build us a city and a tower, whose top may reach unto heaven: and let us make us a name, lest we be scattered abroad upon the face of the whole earth.'"

Genesis 11:4

Since the scattering of the language groups at the place called Babel in the land of Shinar, great civilizations have repeatedly risen at various places around the world. In many cases they have invested tremendous engineering resources and manpower to construct mountainous towers of stone known as pyramids or ziggurats.

Ancient Babylon

The famous city and nation of Babylon we read about in the Bible is in present-day Iraq. The land has a history dating all the way back to the Great Flood.

This ancient baked clay tablet from Babylon (right) records a legendary story of a world-destroying flood that has been linked with the biblical account of Noah's flood.

The Tower of Babel was very likely a pyramid or ziggurat structure, possibly the first one built after the Great Flood. The Genesis record explains that

Model of a Babylonian ziggurat

the purpose of this tower was to "reach unto heaven." This entire culture had abandoned the worship of the Creator who had saved their grandfather Noah just a couple hundred years before. It is important to recognize God's comment about the goal of this culture. He said, "Nothing will be restrained from them which they have imagined to do." Do you really think they imagined they could build a stairway to the stars or the heavenly realm of God? Read the words of Genesis 11 again.

THINK! What are towers typically used for? Are they not the basis for human communication through airwaves? Notice the word "reach" was chosen in Genesis 11:4 to verbalize the action intended by the builders.

When we see how mankind today is preoccupied with "communicating" with heavenly beings, angels, or even "aliens" from outer space, it isn't difficult to understand the inclination of ancient people. They desired interaction with supernatural powers that we would call "demons." And God stopped them short of achieving their goal.

Ancient Egypt

Of the several ancient pyramids found in Egypt today, none is more famous than the Great Pyramid on the plain of Gizeh, across the Nile River west of Cairo. Though generations have been misled to think it was built for a tomb for Pharaoh Cheops, it is now clear that it's purpose was far beyond a mere crypt.

Gizeh is the very center of earth's landmass. All other pyramids are greatly inferior in technical construction. The precision and calculations needed to build the Great Pyramid surpass any project attempted by modern man. It contains more quarried stone than all the cathedrals and churches built in England since the time of Christ. Its builders made a mountain as meticulously as we cut fine gemstones. Many theories attempt to explain the purpose of this incredible structure. Some say Seth built it before the Flood (even if it contains Flood-made rock), but nobody knows who really built it or why. It truly is the greatest manmade wonder of the world.

People are often surprised to learn that Central America has far more pyramids than Egypt.

Ancient Mexico

The magnificent Pyramid of the Sun near Mexico City is an awe-inspiring structure, 216 feet high. This manmade mountain of rock and brick is almost half the height of the Great Pyramid. Its base dimensions of 720 feet by 760 feet almost match the 750-foot exact square of the Gizeh structure.

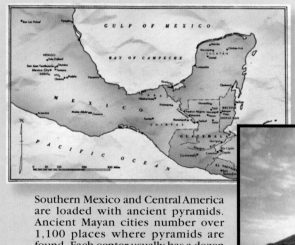

Palenque (above) is another amazing city that displays evidence of a very advanced culture in the middle of a steaming jungle. The tomb of Pacal, an ancient ruler, is strikingly similar in design to the grander structures of Egypt's ancient pharaohs.

The ancient Mayan center of Tikal, in Guatemala, is one of those splendid ruined cities. A dense jungle overtakes it now, and archeologists have exposed only part of it. But it is known to have once covered an area of 50 square miles.

Southern Mexico and Central America are loaded with ancient pyramids. Ancient Mayan cities number over 1,100 places where pyramids are found. Each center usually has a dozen or more pyramids. Some people think there might be over 100,000 pyramids overgrown by the jungles of Mexico.

Teotihuacan, the majestic Mayan city where the Pyramid of the Sun is found, covered eight square miles. Legends say it was built by giant, white demigods. It is an intricately laid-out city with many massive stone buildings. Some 200,000 people are thought to have lived here. This civilization flourished for a thousand years before it fell apart in the eighth century A.D. when the Aztecs came into dominance in that part of the world.

The Pyramid of the Sun

El Tajin is an interesting pyramid because its four sides are covered with little niches. There are 365 of them altogether; one for every day of the year.

Other Mysteries of the Ancient New World

Mayan Astronomy

The ancient Mayans of Central America (c. 2000 B.C.) had a keen awareness of the cosmos. They built observatories like the one at Chichen Itza in the Yucatan Peninsula (right photo). The entire culture of the Mayans seems to have been preoccupied with watching the stars and planets.

Precise mathematics and astronomical calculations enabled the Mayans to figure the length of a solar year to within two ten thousandths of a day. Their year was 365.2420 days long. Our modern astronomers have been able to calculate the solar year is 365.2422 days long.[24]

The ancient inhabitants of Mexico (2000 B.C. to A.D. 900) monitored the orbit of the planet Venus closely. The famous Aztec calendar was specially integrated with the cycles of Venus. There was good reason for this. According to these ancients, the planet Venus was responsible for periodic catastrophes that had previously desolated their society at 52-year cycles. These

astronomical cycles were woven into their religious and cultural patterns.

Our modern intellectual elite tend to dismiss ancient religious practices as mere ritual with no basis in fact. They seem to forget about the law of cause and effect. This is because humanists deny the reality of the spiritual world. Consequently, anything sounding like supernatural judgment coming out of the heavens is ridiculed as unrealistic mythology. The topic of ancient catastrophes destroying entire civilizations is a vivid and genuine reality that deserves further study. Look for further materials from Creation Resource Publications on this subject.

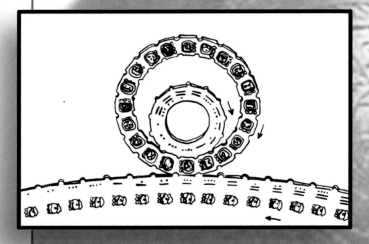

Ancient Mayan astronomers devised an accurate cosmic clock to predict events like solstices and equinoxes (first days of the four seasons).

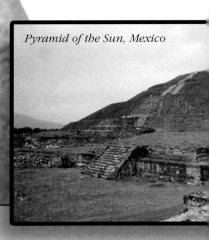

Pyramid of the Sun, Mexico

The Gargantuan Stone Walls of Tiahuanaco

High in the Andes Mountains of Bolivia are the stark and mysterious ruins of Tiahuanaco (dated perhaps as early as 2000 B.C.). This obscure desolation sits at an altitude of 13,000 feet. It seems remarkable that an inhabited city like this could be positioned where the air is so thin. Yet this mysterious city may help us to better appreciate the abilities of the ancients.

Almost nothing is known of the history of this place. Who lived there? When did it flourish? How far did its empire extend?

In the book, *The World's Last Mysteries*, by Reader's Digest, the lead headline reads: *"When the winds of the Andes howl through Tiahuanaco's deserted buildings, it is easy to believe in the Indian legend that the city was built by a race of giants."*

At Tiahuanaco there are massive stone stairways that defy simple explanation. They seem to have been built specially to accommodate a race of large people. There are sections of gigantic stone walls that have been hurled to the ground by what must have been a spectacular earthquake. Wall sections more than a foot thick and 16 by 26 feet in size, have been strewn around like so many toy blocks. What could do this?

A Jesuit priest who apparently interviewed native people of the area in the days of the Spanish conquistadors wrote this amazing account: *"The great stones one sees at Tiahuanaco were carried through the air to the sound of a trumpet."*[25]

That may sound far-fetched in terms of our experience, but does that mean such a technique is absolutely impossible? The writer of the article in the previously mentioned book must think so. He discounts the priest's early report by writing, *"There must have been people even then who were not satisfied with an explanation that invoked the use of magic."*

THINK! A hundred years ago, if you tried to explain to a person living then how a Boeing 747 takes off from the runway, you might also be accused of hallucinating or dabbling in witchcraft.

Is there a reasonable explanation?

The enormous "Gateway to the Sun" is hewn from a single rock and weighs as much as 10 tons. Other single cut stones here weigh up to 100 tons! Like the other volcanic building stones of this center, it was moved from a quarry at least 60 miles away, across rough terrain without wheels and roads! How did they do it?

How Can Massive Stone Blocks Be Moved?

The Ancient Ruins of Baalbek, Lebanon

The prominent 70-foot-high columns seen at Baalbek date back to Roman times. These imposing pillars are amazing. They were quarried in Egypt, floated across the Mediterranean Sea and then hauled across the mountains before being erected here. The engineering to accomplish that feat alone is nothing short of a marvel.

But a glimpse of the foundation stones on which the columns stand is even more surprising. See the tourists walking on it? Ancient people before the Romans are responsible for cutting and moving these gigantic blocks from a nearby quarry. The size of these carefully hewn blocks truly boggles the mind. One course of cut stones set 26 feet high in a wall here contains several of the largest building blocks ever used in construction. They are each over 60 feet long and fitted so perfectly together that you can't even get a razor edge between them!

In the nearby quarry, the builders of this amazing center left behind a masterfully cut stone. It measures 14 feet high, 13 feet wide and 70 feet long! It weighs at least **a thousand tons!** How these blocks were transported, raised and fitted so perfectly is a baffling mystery to modern engineers. No modern equipment or techniques can move these blocks like that.

How do modern engineers move huge blocks of stone?

The *National Geographic* magazine dated May, 1969 featured one of the most ambitious engineering and preservation projects of all time. The great Egyptian monument at Abu Simbel, along the Nile River, had to be moved to avoid being submerged by the rising waters of the newly formed reservoir behind the Aswan Dam called Lake Nasser.

The 3,200 year old monument is one of the most striking examples of ancient Egyptian engineering and art. It was carved into natural rock cliffs beside the Nile River. To move it, modern engineers had to cut the towering 67 foot high temple into over 1,000 pieces of 20 to 30 tons each.

By the end of 1969, after four and a half years of work and an expense of 40 million dollars, the project was completed. The great temple was raised 200 feet to higher ground. Despite our modern machinery, one can't help asking:"Did the ancients have a simpler and more ingenious way of doing it?"

The magnificent Roman pillars in the background of this page are in Baalbek, Lebanon. They have been called "mysteries" by *National Geographic* magazine (April 1958).

An Amazing Report from Tibet

In the book by authors Playfair and Hill, *The Cycles of Heaven*, a report was included from Tibet. In the 1950s, a Swedish aircraft engineer named Henry Kjellson is reported to have witnessed an amazing ceremony conducted by Tibetan priests. At the base of a sheer rock cliff in the mountains, scores of men had gathered to take their parts in the dramatic scene. Groups of them were carefully arranged in a semi-circle, equipped with large suspended drums and others with special trumpets.

As the ceremony proceeded, the drum beats and trumpet blasts were directed at the center of the semi-circle in front of the cliff. A four-foot block of rock was positioned there. The corporate noise of the assembled instruments must have been deafening. But after a while the heavy chunk of rock (weighing several tons) was seen ascending in the air straight up to the top of the cliff.

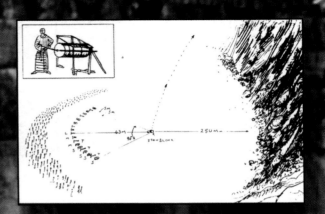

We tend to think of a report like that as either some sort of witchcraft or just plain false. It just sounds too bizarre to be explainable by natural laws, but is such an action possible?

Notice a report from the popular *OMNI* magazine in November 1980. According to the report, **NASA scientists have succeeded in using sound waves to levitate pellets of glass or metal!**

"There is no remembrance of former things. There is nothing new under the sun. It has already been in ancient times before us."

Solomon, 1,000 B.C.

A portion of the great monument of Pharoah Ramses at Abu Simbel along the River Nile in Egypt

Think ! Could there be realms of technology which we have yet to re-discover? Do you recall Solomon's words?

China's Incredible Find

The April 1978 issue of *National Geographic* magazine featured one of the most amazing discoveries of modern archeology. Hidden for years in legend, the report says the tomb of China's first emperor has finally been found!

The excavation will take years to complete. The total enclosure is over 500 acres! It is called the Spirit City of Ch'in Shih Huang Ti. It dates back to 210 B.C., over 2,200 years ago.

The evidence of skilled technology and accomplishment is abundant. Imagine the kilns needed to fire thousands of life-size horses! They are all lined up in battle marching formation – an immense ceremonial parade to escort the emperor into eternity. The whole assembly is completely roofed over and buried. Much of ancient Chinese history will no doubt be filled in as this important discovery is more thoroughly explored.

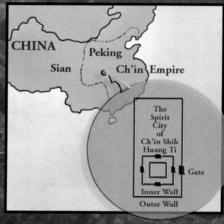

Every soldier's face is sculpted with a unique expression.

A virtual terra cotta (clay) army, the elaborate burial includes over 6,000 life-size pottery men and horses too!

How advanced were the ancient orientals?

The personal crypt of China's first emperor has yet to be unearthed here at the Spirit City. His accomplishments give some indication of the capability of ancient men in the Far East. He was very progressive and the first to successfully unify the country. He destroyed the ancient feudal system and centralized the empire.

It was this man, beginning his reign at age 13, who completed the Great Wall of China (largest manmade structure on Earth). It's over 1,500 miles long. With 25,000 watchtowers, the 20-foot high wall supported a roadway that allows six horsemen abreast to march on its pavement.

It was Ch'in Shih Huang Ti who assembled China's first standing army. It may have numbered several million men altogether.

He also codified China's laws, standardized the system of Chinese writing, and built a vast network of public roads and canals.

These accomplishments give us only a little insight into the fabulous ingenuity of mankind. Here in China, as in other ancient centers, we see the tendency of men to create great projects and make a name for themselves as they dominate people.

Endless Discoveries Continue to Reveal the Truth about Ancient Man

More Advancements from Ancient China

Another remarkable burial, this time of a noble-woman, was featured in the May 1974 issue of *National Geographic* magazine as "A Lady from China's Past." The elaborate burial was like an underground building, carefully engineered with protective layers of charcoal and white clay. Several ornately decorated caskets were found, each inside the other.

How did they preserve this corpse so well 2,100 years ago? Advanced knowledge would have been necessary to halt natural decay processes so immediately and for such a long time. In our modern age we are just beginning to understand such preservation by observing what the ancients did.

Found in the tomb with this woman's body were a number of other skillfully made artifacts. Each of them further opens our understanding of the fact that these ancient people were quite technologically advanced in many ways.

An ornately decorated cosmetics case accompanied this high-class socialite. In the case was a silk scarf and other 20th century conveniences like mitts, hairpiece, face powder, rouge, comb and brushes.

Also found in the tomb were 180 pieces of exquisite lacquerware jars and table service items. These lightweight containers are highly crafted and artistically decorated. In ancient China these acid-resistant vessels were esteemed ten times more costly than similar bronze objects.

Even more surprising was the discovery of printed cloth! Fine silks numbering some 50 lengths, comprise the most lavish cache of ancient fabrics ever found. When you look closely at the intricate repeated designs on the cloth you realize they are not painted on; they are printed on. And keep in mind that this was done over 2,100 years ago. Maybe the printing craft was just re-invented by Johann Gutenburg in Germany about 1450. Again, this is just another indication of the truth of Solomon's words of 3,000 years ago: "There is nothing new under the sun."

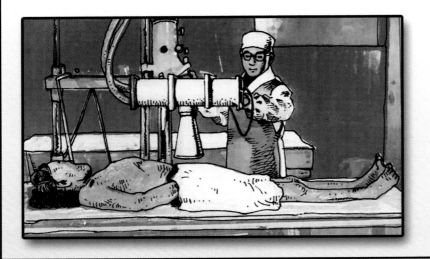

When the Chinese specialists opened the innermost casket they were in for the surprise of their lives. Under 20 layers of fine silken wrappings, the astonished scientists discovered the woman's body was almost perfectly preserved! Her skin is still pliable. Her hair is still firmly rooted in her scalp. Modern X-ray technicians can even identify her internal organs, and the menu of her last meal eaten the day she died!

Puzzling Artifacts or Expected Technology?

In modern times, dentists have learned to use the lost wax method for making tooth crowns. But man-made tooth crowns were also fashioned by Aztec dentists of ancient Mexico!

Crystals in Ancient Communication?

Not so long ago, in the days of great grandfather (the early 20th century) it was common for a young boy to make and use a simple "crystal radio set." All it takes is a rock crystal, a wire and an earpiece. Then you can listen to radio broadcasts transmitted from miles away through invisible airwaves. We think of that as a modern discovery, but is it? God made crystals with the unusual property of being useful in the transmission and reception of radio waves. Much of modern communication technology is dependent on the use of microcrystals in elaborate electronic gadgetry. So why should we insist we are the first to discover this miracle of science?

A number of mysterious indications from ancient cultures link the use of rock crystals with communications through the air. Even the Bible draws this connection. Why were crystals specifically part of the breastplate prescribed by God to be worn by the high priest when using the mysterious "urim" and "thummim" to find the will of God in a matter? God designed the sensory systems of our bodies with

Some ancient human skulls reveal that cranial surgery was performed with at least some success on the Inca people of ancient Central America. The discovery of many skulls like this show that the bone tissue partially refilled the surgical removal through normal healing processes.

integrated electrical circuitry. Like so many discoveries of science and technology, we create inventions by observing God's creation. So why shouldn't we learn even more about this by applying the wisdom of Solomon regarding the ancients? Since far more information about ancient technology has been lost than has been recovered, it should be no surprise if we someday learn that even the realm of wireless communications was known by people before and after the Great Flood.

Traditions of the Mayans of Central America and Mexico insist that phenomenal things were done in the realm of communications. A 1,000-pound slab of solid crystalline rock rests in a special high tower in the mysterious ancient city of Palenque. The priestly rulers of this vanished civilization are said to have come here to transmit and receive audible messages from great distances away.

Nothing New Under the Sun?

As you continue to read reports of astonishing finds produced by
ancient civilizations, the key to the mystery becomes clearer.
Remember, "Prepare yourself" as you "enquire of the former age
and consider the accomplishments of the ancients."

More Than a Ceramic Jar

In 1957, a strange relic was found in a Baghdad museum. It was
excavated in 1936 but it was made by the Parthians, who ruled that
region of Iraq from 250 B.C. to A.D. 224. Suspended in a simple
earthenware vase was a copper cylinder with an iron rod inside.
Its parts were cemented with asphaltum (the bitumen of Genesis)
and soldered with a 60/40 tin/lead alloy just like what is used today.
When a General Electric engineer made a replica of the object and
added grape juice, it produced a half-volt of electricity! He even
used the current to electroplate a silver statuette with gold. He
believes many of the "gold" artifacts from ancient time are really
made of silver and plated with gold. But it wasn't until 1799 that
Mr. Volta "rediscovered" how to make an electrical storage battery.[26]

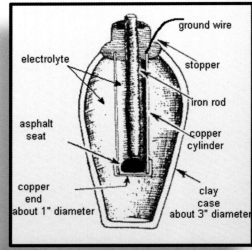

Computers 2,000 Years Ago?

A Greek merchant ship sank off the
coast of the tiny Mediterranean
island of Antikythera about 65 B.C.
The first underwater archeological
mission brought up its treasures in
A.D. 1900. But it wasn't until 1951
that a Yale University professor
painstakingly revealed that a Greek
technician had made a complex
mechanical computer over 2,000
years ago. With timepiece precision,
his accurately fashioned instrument
used some 20 bronze gears to
compute the motions of the sun,
moon and planets. A wooden box

the size of a large book contained what has now become famous as "the Antikythera Mechanism." Engraved brass instruction plates, dials, and integrated wheels baffled historians who assumed that the differential gear concept was not invented until the 17th century A.D.[27]

A Strange Little Wooden "Bird?"

Rummaging through some boxes of discarded exhibits in the Cairo Museum of Antiquities, an Egyptologist in 1969 noticed a box marked, "wooden bird models." Puzzling over one particular small model, his 20th century mind realized this was not a carving of a bird. It had a slightly curved overhead wing, a sleek, tapered body, and a vertical tail fin. It was a carefully proportioned, one-ounce scale model of an aircraft called a pusher-glider.

A committee of scientists appointed by Egypt's Ministry of Culture concluded that the seven-inch wingspan model of light sycamore wood, incorporated design concepts that have since taken modern aviation engineers decades of experimentation to perfect. The proportions are so exact that a full-sized plane of this design with minimal engine power could stay airborne at speeds as low as 45 miles per hour while carrying an enormous payload. With a wing design like the Concorde super jet, this 2,000-year-old relic makes us realize that its maker was not unaware of sophisticated aerodynamics. Numerous other artifacts and images from various ancient cultures suggest the same reality. So, how far did the people of the "former age" go in their pursuit of flying machines?[28]

From our brief survey of ancient advancements it is clear that the popular evolutionary premise has seriously misled us. The most ancient people on Earth were ingenious. None of them recall a backward beginning. They commonly recall the Great Flood. As far as true science is concerned, man has always been 100 percent human! And why is man so capable? No one has declared the answer more profoundly than King David of 3,000 years ago:

You [Almighty God] have made man a little lower than the angels, and You have crowned him with glory and honor.
Psalm 8:5

Where does the truth about mankind's past lead us?

Come See the Awesome Works of God

> "Come and see the works of God. He is awesome in his doing toward the children of men."
>
> *Psalm 66:5*

> "I will remember the works of the Lord.... I will meditate also of all thy work, and talk of thy doings....Thou art the God that does wonders. Who is so great a God as our God?"
>
> *Psalm 77:11, 12, 14*

> "One generation shall praise thy works to another, and shall declare thy mighty acts.... The Lord is good to all and his tender mercies are over all his works.... You [Lord] open your hand and satisfy the desire of every living thing.... The Lord is near to all them that call upon him, to all that call upon him in truth."
>
> *Psalm 145:4, 9, 16, 18*

The Purpose in Creation and the Lesson of History

The more we discover the world around us, the more we're impressed by the purposes of an Awesome Designer-Creator. But the most important purpose is the Designer's intention that we who are made in His image would give Him the credit and seek to please Him. A good prayer is: "Lord, open my eyes to appreciate your marvelous works all around me!" When I do that, a sense of awe fills my thinking.

"Awesome are Thy works, O Lord!" That's a genuine and personal confession (Psalm 66:1). Somehow, when I recognize God's greatness in His creation, and freely thank Him for it, a deep faith rises in my heart. Confidence in God's overarching providence calms my anxieties about temporal conditions. Like Jeremiah (Jer. 32:17), I will confess: "Ah Lord God! Behold you have made the heavens and the earth by your great power and outstretched arm. There is nothing too hard for you!"

Similarly, we see how those who try to ignore their Creator or His laws become so arrogant that they set themselves up as supreme authorities. They fashion themselves into humanistic gods. And because the Creator made man with a core spiritual nature, a nature that desires to worship, it's not surprising to see whole societies follow intimidating conquerors, Caesars, cult leaders, politicians and modern celebrities who promote

selfish gratification of natural desires, and the denial of the true living God.

Regardless of our ideas about God, He always gives us freedom of choice. We can seek the Truth found in the Bible, or we can make excuses why we don't believe it, even if we refuse to "prove all things" and "reason" with God.

Can the Creator's Laws Be Broken?

Isn't it amazing how simple solutions defeat grave problems? Yet how sad to see the ignorant suffer (or those who reject the solutions). Simple vitamin C prevents and cures the dreaded scurvy and other diseases. Consider that simply washing hands reversed the high death rate of patients in 19th century hospitals. Now we're surprised that deadly diseases are preventable by reducing toxic chemicals in our environment and increasing our consumption of natural raw foods. The law of cause and effect is unapologetically consistent. Ignorance of the law doesn't excuse anyone from being subject to it.

Everyone knows deep down that he has disobeyed God's laws. We've cheated,

stolen, lied, lusted, and followed false gods of our own making.... We might fool ourselves into thinking we're just as good as the next guy, but in reality we know we're just as selfish and sinful. We know the Creator's indisputable laws condemn us. We know that we reap what we sow, but so many try to ignore that law. Then we try to justify ourselves and hope for divine acceptance, based on our relative goodness. We know judgment awaits us all. That's why so many fear death. There is no peace for the wicked, and the surprise of condemnation awaits the self-righteous. However, the honorable Judge of all offers mercy and forgiveness to all who hunger for it. Faith in **what He did** is His perfect solution to our hopeless separation from Him.

Regardless what you believe, are you aware of your own capacity to be deceived?

Do you truly want to be **free** from every intellectual, emotional and spiritual trap that would enslave you to false gods and empty promises? Jesus said in John 8:31-32:

> *"If you continue in my word, then are you my disciples indeed; and you shall know the truth, and the truth shall make you free!"*

If truth is the **key** to genuine personal liberty, then **who** has the truth? Jesus made no apology for plainly stating that **He is** the Truth!! Jesus said in John 14:6: "I am the way, the truth and the life: no man comes unto the Father, but by me." No ordinary man or even a prophet could say **that** and be considered sane. That's why **we need to determine FOR OURSELVES that Jesus is who the Bible claims He is — the very Creator of the universe!** Then we must admit that we've ignored Him, gone our own way, and missed His mark of a God-honoring life. When we confess by faith in His Word, that Jesus Christ has bought us with His blood, and given us eternal life by His resurrection, then His Word gives us confidence, and His Spirit confirms to our hearts, that we have truly been accepted by God. Jesus was not the redeemer of a few, but of all mankind. Next question?

If you want to know the truth...

...then are you willing to "continue" to live in the light of God's Word and **be a true disciple of Jesus in action?** Paul the Apostle of Jesus Christ warns that in the last days people would be **"ever learning but never able to come to the knowledge of the truth."**

If we set our affection on knowledge alone, we will miss the Author of knowledge himself!

Jesus taught to set our personal affection on Him so His Holy Spirit will lead us personally to the knowledge of the truth. Like the Psalmist, I must frequently refresh my relationship with Him and ask: **"Search me, O God, and know my heart: try me, and know my thoughts: and see if there be any wicked way in me, and lead me in the way everlasting"** (Psalm 139:23-24).

Ignorance that destroys is basically linked to the willful rejection of God's law.

> *My people are destroyed for lack of knowledge; because you have rejected knowledge, I will also reject you... seeing you have forgotten the law of your God, I will also forget your children.*　　Hosea 4:6

It's time to reject ignorance, admit our disobedience, and embrace God's mercy.

> *Seek ye the Lord while he may be found, call ye upon him while he is near; Let the wicked forsake his way and the unrighteous man his thoughts and let him return unto the Lord, and he will have mercy upon him: and to our God, for he will abundantly pardon. "For my thoughts are not your thoughts, neither are your ways my ways," says the Lord. "For as the heavens are higher than the earth, so are my ways higher than your ways, and my thoughts than your thoughts. For as the rain comes down, and the snow from heaven, and returns not there, but waters the earth, and makes it bring forth and bud, that it may give seed to the sower, and bread to the eater: So shall my word be that goes forth out of my mouth; it shall not return unto me void, but it shall accomplish that which I please, and it shall prosper in the thing whereto I sent it."*　　Isaiah 55:6-11

The Next Step?

Knowing there are many untrustworthy and Godless influences in the world, what must be personally taken to heart about the knowledge of creation?

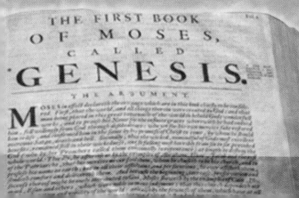

> *We will not hide them from their children, showing to the generation to come the praises of the Lord, and his strength, and his wonderful works that he hath done... that the generation to come might know them, even the children which should be born; who should arise and declare them to their children; that they might set their hope in God, and not forget the works of God, but keep his commandments.*　　Psalm 78:4, 6, 7

References

Section 1
Unlocking the Mysteries of the Early Earth

1. *Discover,* (Oct. 1985).

2. *Time,* (May 1, 1989).

3. *National Geographic,* (May, 1974), p. 625.

4. Cyr, Donald L., "The Crystal Veil," *Stonehenge Viewpoint,* Issue 106, Santa Barbara, (1995).

5. Dillow, Joseph C., *The Waters Above,* (Chicago, Ill.: Moody Press, 1982), p. 58.

6. Patten, Donald, *The Biblical Flood and the Ice Epoch,* (Seattle, Wash.: Pacific Meridian Pub., 1966), p. 194-224.

7. D. G Whitley, "The Ivory Islands in the Arctic Ocean," *Journal of the Philosophical Society of Great Britain,* XII (1910): p. 49 in *Earth in Upheaval,* by I. Velikovsky, (Dell ed., 1955), p. 19.

8. Thomas G. Barnes, *Origin and Destiny of the Earth's Magnetic Field,* Technical Monograph #4, (El Cajon, Calif.: Institute for Creation Research, 1983).

9. The science journal *Nature,* reported May 13, 1999 on the Reuters news service, disclosed that "scientists had thought gigantism... was a phenomenon limited to a few species. But Belgium based Gauthier Chapelle and Lloyd Peck have shown size is dependent on oxygen availability, which is greater in cooler and less saline locations like the Antarctic Ocean."

10. Dixy Lee Ray, *Environmental Overkill*: whatever happened to common sense? (New York, N.Y.: Harper Perennial, 1994).

11. *Bible Science Newsletter,* (January, 1975), p. 6.

12. Henry Morris, *Scientific Creationism,* (San Diego, Calif.: Master Books, 1974), p. 152.

13. Melvin Cook, *Prehistory and Earth Model,* (London, England: Max Parrish, 1966).

14. Hans Pettersson, "Cosmic Spherules and Meteoritic Dust," *Scientific American,* vol. 202 (2/1960), p. 132, in *Scientific Creationism* by Henry Morris, p. 152; Also see Fletcher G. Watson, *Between the Planets* (Cambridge, Mass.: Harvard Univ. Press, 1956) (estimates accumulation of 300-thousand to 3-million tons per year deposited on the earth).

15. Morris, *Scientific Creationism,* p. 154.

16. Franklyn M. Branley, *Apollo and the Moon* (Garden City, NJ.: Natural History Press, 1964), (published for American Museum/Hayden Planetarium -- "Some astronomers think that in places, lunar meteoritic dust may be a hundred feet or more deep. Also, it may be so loosely packed that a spaceship would sink into it, never to be seen again."). Also see *The Rand McNally New Concise Atlas of the Universe* (London, England: Mitchell Beazley Publishers, 1978), p. 41: "The theory that the Maria were covered with deep layers of soft dust was current until well into the 1960s."

17. Morris, *Scientific Creationism,* p. 155.

18. Edward Blick, *Correlation of the Bible and Science,* (Oklahoma City, Okla.: Southwest Radio Church, 1976), p. 28.

19. Benjamin Allen, "The Geological Age of the Mississippi River," *Creation Research Society Quarterly,* vol. 9 (9/72), p. 96-114.

20. Patten, *The Biblical Flood...,* p11.

21. John Whitcomb and Henry Morris, *The Genesis Flood,* (Philadelphia, Penn.: The Presbyterian and Reformed Publishing Co., 1961), p. 408-409.

22. Morris, *Scientific Creationism,* p. 156.

23. John C. Whitcomb, Jr., *The World that Perished,* (Grand Rapids, Mich.: Baker Book House, 1973), p. 114.

24. Whitcomb and Morris, *The Genesis Flood,* p. 392.

25. Morris, Scientific Creationism, p. 167-169.

26. Thomas, G. Barnes, *Origin and Destiny of the Earth's Magnetic Field,* (San Diego, Calif.: Institute for Creation Research, 1973), in Morris, *Scientific Creationism,* p. 157.

27. Coe, R.S., Prevot, M., and Camps, P., "New evidence for extraordinarily rapid change of the geomagnetic field during a reversal," *Nature,* 374 (20 April 1995). And Austin, S.A., Baumgardner, J.R. Humphreys, D.R., Snelling, A.A., Vardiman, L., Wise, K.P., "Catastrophic Plate Tectonics: A Global Flood Model of Earth History," *Proceedings of the Third International Conference on Creationism,* 1994, p. 609.

28. Overn, Wm., "The Tilt of The Earth's Axis, Its Orientation Brings The New Age, Its History Reveals The Flood and Explains The Magnetic Reversals," *Proceedings of the 1992 Twin-Cities Creation Conference,* p. 84. and

    Setterfield, B., "The recent change in the earth's axis", *Science At The Crossroads,* 1983 National Creation Conference, Minneapolis, MN, p. 82-84.

29. Morris, *Scientific Creationism,* p. 153.

30. Melvin Cook, "Where is the Earth's Radiogenic Helium?" *Nature,* v. 179, (1/26/57), p. 213, in Blick, p. 26.

31. Jonathan Henry, *The Astronomy Book,* (Green Forest, Ark.: Master Books, 1999), p. 39.

32.1 'R.R. Britt, "New Photos of the Sun are Most Detailed Ever," Space.com, 13 Nov. 2002'

32.2 Russell Akridge, "The Sun is Shrinking," *Acts and Facts Impact* #82, Institute for Creation Research, San Diego, (4/80).

33. Halton Arp, *Quasars, Redshifts and Controversies,* (Interstellar Media, 1987).

34. Keith Davies, "Distribution of Supernova Remnants in the Galaxy," *Proceedings of 3rd Int'l Conf. On Creationism,* (1994): p. 175-182, Creation Science Fellowship, Inc., Pittsburgh, USA.

35. D. Goldsmith, "Digging deeply in galaxies' pasts," *Science* 271:450 (1966).

36. Keith Davies (quote), in *This Week in Bible Prophecy* TV program 191 (1995), Lalonde & Lalonde Creative Inc., Canada.

37. Tom Van Flandern, *Dark Matter Missing Planets & New Comets,* p 141.

38. John C. Whitcomb Jr., *The Origin of the Solar System,* (Philadelphia, Penn.: Presbyterian and Reformed Publishing Co., 1976).

39. *Science Magazine,* vol. 167, (1/30/70).

40.1 Morris, *Scientific Creationism,* p. 147.

40.2 Robert Kofahl, *Handy Dandy Evolution Refuter,* (San Diego, Calif.: Beta Books, 1977) p. 115 citing Cherdyntsev, V.V., et al., Geolog. Insti. Academy of Sciences, USSR, Earth Science Section, 172, p. 178. The data is reproduced by Sidney P. Clementson in Creation Research Society Quart. 7, Dec. 1970, p. 140.

41. *Radiocarbon* (journal), vol. 11, (1969).

42. Robert Whitelaw, "Time, Life, and History in the Light of 15,000 Radiocarbon Dates," in *Speak to the Earth,* edited by George Howe, (Philadelphia, Penn.: Presbyterian and Reformed Publishing Co., 1975), p. 339.

43. Glen S. McLean, personal interview in Eston, Sask., Canada in 1984.

44. W.F. Libby, *Radiocarbon Dating* (Chicago, Ill.: Univ. of Chicago Press, 1955), p. 7, as in Morris, *Scientific Creationism*, p. 164.

45. Robert Whitelaw, "Time, Life, and History in the Light of 15,000 Radiocarbon Dates," in *Speak to the Earth*, p. 339.

46. Kofahl, *Handy Dandy Evolution Refuter*, p. 119, citing others.

47. Bruno Huber, "Recording Gaseous Exchange Under Field Conditions", *The Physiology of Forest Trees*, K. V. Thinmann ed., (New York, 1958) p. 194.

48. Robert E. Lee, "Radiocarbon, Ages in Error," *Anthropological Journal of Canada*, vol. 19, no. 3, (1981), p. 9-29.

49. William D. Stansfield, PhD., (animal breeding instructor of biology, California Polytechnic State University) in *The Science of Evolution*, (New York, N.Y.: Macmillan, 1977), p. 84.

50. Curt Teichert, *Bulletin* of the Geological Society of America, vol. 69, (January 1958).

51. Note: Was the human reproduction fertility cycle different when first created? Genesis 3:16 has some bearing here. We see that some large animals are able to conceive only several times a year. With a projected life span of hundreds of years, the human cycle might reasonably have been programmed for a year or more, especially since it appears that children did not reach adolescence at as early an age as we see now.

Section 2

Unlocking the Mysteries of Evolution

1. From a letter to a Mr. Grose, signed by Dr. Werner von Braun, read by Dr. John Ford to the California State Board of Education on 9/14/72 and cited in *Jesus Christ Creator* by Kelly Seagraves 1973, and later printed in *Applied Christianity*, then quoted in the *Bible Science Newsletter*, (May 1974), p. 8.

2. G. Richard Bozarth, "The Meaning of Evolution," *American Atheist*, (Feb. 1978), p. 30.

3. We quote from the 1990 University of Chicago facsimile edition p. xvii of Lyell's *Principles of Geology*, where this quote is referenced as "Lyell 1881 vol. 1. p. 268." This does not, however, refer to the 1881 edition of Lyell's *Principles of Geology* but to his *Life Letters and Journals*, published by John Murray 1881. Thanks to Dr. Kohl for tracking this down.

4. Richard Dawkins, "What's All the fuss About?" *Nature*, vol. 316 (Aug. 22, 1985), p. 683.

5. Julian Huxley, "Evolution and Genetics," in James Roy Newman, editor, *What is Science*? (New York, N.Y.: Simon and Schuster, Inc., 1955), p. 278.

6. Richard Lewontin, "Billions and Billions of Demons," *The New York Review*, (Jan. 1997), p. 31. The complete quote: "We take the side of science *in spite of* the patent absurdity of some of its constructs, *in spite of* its failure to fulfill many of its extravagant promises of health and life, *in spite of* the tolerance of the scientific community for unsubstantiated 'Just So' stories, because we have a prior commitment, a commitment to materialism. It is not that the methods and institutions of science somehow compel us to accept a material explanation of the phenomenal world, but, on the contrary, that we are forced by our a priori adherence to material causes to create an apparatus of investigation and a set of concepts that produce material explanations, no matter how counterintuitive, how mystifying to the uninitiated. Moreover, that materialism is absolute, for we cannot allow a Divine foot in the door."

7. Charles Darwin, in a book by Elbert Hubbard (of Roycrofters fame) called *Little Journeys to the homes of Great Scientists*, publ. by William H. Wise, 1916.

8. George Gaylord Simpson, *Science* (1964) 143:769.

9. "In the (Van Flandern model), the required finite range for the force of gravity explains the behavior of galaxies and eliminates the need to hypothesize dark matter." Tom Van Flandern, *Dark Matter, Missing Planets and New Comets*, Glossary, (North Atlantic Books, 1993), p. 393.

10. *National Geographic*, (June 1983), page 741.

11. *The Globe and Mail*, October 24, 1998, p. D5, by Gregg Easterbrook, based on his book, *Beside Still Waters: Searching for Meaning in an Age of Doubt*.

12. George Wald, "The Origin of Life," in *Scientific American*, vol. 191(2), (Aug. 1954), p. 48.

13. John Eddy, PhD, quoted in a report by R.G. Kazman, "It's About Time: 4.5 billion Years." (a report on a symposium at the Louisiana State University), in *Geotimes*, vol. 23, September 1978, p. 18.

14. Michael Denton (molecular biologist), *Evolution: A Theory in Crisis* (Bethesda, Md.: Adler and Adler, 1986), p. 264.

15. I. L. Cohen, mathematician, researcher, author, member of New York Academy of Sciences, and officer of the Archaeological Institute of America, *Darwin Was Wrong; A Study in Probabilities* (New Research Publications, Inc., 1984), p. 209.

16. Gary Parker and Henry Morris, *What is Creation Science* (El Cajon, Calif.: Master Books, 1982), p. 40.

17. David Green and, Robert F. Goldberger, *Molecular Insights into the Living Process*, (New York, N.Y.: Academic Press, 1967), p. 406-407.

18. Loren Eiseley, Ph.D., *Anthropology: The Immense Journey* (New York, N.Y.: Random House, 1957), p. 199.

19. George Wald, "The Origin of Life," *Scientific American*, 191:48 (May 1954).

20. Denton, *Evolution: A Theory in Crisis*, p. 338.

21. Charles Darwin, "*The Origin of Species…*," 1859, (in a modern undated edition), p. 133.

22. Sir Fred Hoyle, in "Hoyle on Evolution," in *Nature*, vol. 294, (11/12/1981), p. 105.

23. Ernst Mayr, *Populations, Species and Evolution* (Cambridge: Harvard Univ. Press, 1970), p. 102.

24. Pierre-Paul Grasse, *Evolution of Living Organisms* (New York, N.Y.: Academic press, 1977), p. 88, 103.

25. W.R. Thompson, *Introduction to the Origin of Species*, by Charles Darwin (N.Y., N.Y.: Dutton, 1956).

26. Though the source of this quote has been lost, it serves well to visualize the futility of evolutionary dreaming.

27. Majerus, *Melanism*: "Evolution in Action," reviewed by Jerry A. Cohen in Nature (Vol. 396, p. 35).

28. Charles Darwin, *The Origin of Species*, Sixth Edition (New York, N.Y.: The Modern Library, 1872), p. 66.

29. *Science Digest* magazine, (Nov-Dec. 1980), p. 25.

30. David Raup, *Bulletin of the Field Museum*, vol. 50(1), (1979), p. 22-29.

31. Charles Darwin, *Origin of Species*, (many editions) Chapter VI, Difficulties of the Theory, and Chapter X, On the Imperfection of the Geologic Record.

32. Steven M. Stanley, *Macroevolution: Pattern and Process* (San Francisco, Calif.: W. H. Freeman and Co., 1979), p. 39.

33. Stephen Jay Gould, "Evolution's Erratic Pace," *Natural History*, vol. 86(5), (1977), page 14.

34. Stephen Jay Gould, "Is a New and General Theory of Evolution Emerging?" *Paleobiology*, vol. 6 (Winter 1980), p. 127.

35. R.R. Rastall, wrote in *Encyclopaedia Britannica*, vol. 10 (1956), p. 168.

36. Derek Ager, "Fossil Frustrations," *New Scientist*, vol. 100, no. 1383 (Nov. 10, 1983), p. 425.

37. *Vancouver Sun*, Canada (Oct. 9, 1980).

38. H.S. Ladd, Geologist, *Geological Society of America Memoir* 67 (1957).

39. George G. Simpson, *The Meaning of Evolution*, (New Haven, Conn.: Yale Univ. Press, 1949), p. 18.

40. Francis Crick, "*Life Itself – Its Origin and Nature*," (London, England: Futura, 1982), quoted in Mark Eastman, and Chuck Missler, "*The Creator Beyond Time and Space*," (TWFT Publishers, 1996), p. 62.

41. quoted in *Bible Science Newsletter*, (May 1974), p. 6.

42. Sharon Begley, "Science Contra Darwin," *Newsweek* (April 8, 1985), p. 80.

43. D.M.S. Watson, in *Nature*, vol. 124, p. 233, (1929).

44. Robert Jastrow, *God and The Astronomers*, p. 116.

45. Denton, *Evolution: A Theory in Crisis*, p. 340-341.

Section 3
Unlocking the Mysteries of Original Man

1. *The Last Two Million Years*, (New York, N.Y.: Reader's Digest Assoc., 1973), p. 12.

2. *Science Digest*, (April 1981), p. 36.

3. Marvin L. Lubenow, *Bones of Contention*, (Grand Rapids, Mich.: Baker Book House, 1992), p. 163.

4. ibid.

5. See the drawing on p. 91 of James Perloff's excellent book, *Tornado In A Junkyard*, Arlington, Mass.: Refuge Books, 1999).

6. Lubenow, p. 114-119.

7. Stephen Jay Gould, *Eight Little Piggies* (New York, N.Y.: W.W. Norton, 1993), p. 135, in Perloff's *Tornado in a Junkyard*, p. 85.

8. Ian T. Taylor, *In the Minds of Men: Darwin and the New World Order* (Minneapolis, Minn.: TFE Publishing, 1991), p. 238-239. and *Ape Men: Fact or Fallacy* by Malcolm Bowden (Sovereign Pub., 1977)

9. Lubenow, p. 60.

10. *Sacramento Union*, (Sept. 16, 1981).

11. *Smithsonian* magazine, (October 1986).

12. Perloff, p. 82.

13. Taylor, p. 232.

14. Mark Twain, *Life on the Mississippi*, p. 156.

15. Philips Verner Bradford, Harvey Blume, *Ota Benga: The Pygmy in The Zoo* (New York, N.Y.: Delta Books, 1992).

16. Taylor, p. 204-206.

17. Richard Dawkins, Oxford wrote: "…there are certain things about the fossil record that any evolutionist should expect to be true. We should be very surprised, for example, to find fossil humans appearing in the record before mammals are supposed to have evolved! If a single, well-verified mammal skull were to turn up in 500 million year old rocks, our whole modern theory of evolution would be utterly destroyed. Incidentally, this is a sufficient answer to the canard, put about by creationists and their journalistic fellow travelers, that the whole theory of evolution is an 'unfalsifiable' tautology. Ironically, it is also the reason why creationists are so keen on the fake human footprints, which were carved during the depression to fool tourists, in the dinosaur beds of Texas," (*The Blind Watchmaker*, 1986, p. 225).

18. F.A. Barnes., "The Case of the Bones in Stone," *Desert Magazine*, v. 38 (Feb. 1975) p. 36-39, and Mysteries of the Unexplained, (New York, N.Y.: Reader's Digest, 1982), p. 41.

19. Don R. Patton, offers further information at the website www.bible.ca/tracks.

20. An articulated fossil skeleton is together, connected in the rock as in life. When a cow dies on the open range, a few weeks later the animal is decayed and the bones are scattered by scavengers. Even individual bones rarely become fossils because small rodents and insects reduce them to powder. "Articulated" skeletons obviously indicate rapid burial. "In other formations where articulated skeletons of large animals are preserved, the sediment must have covered them within a few days at the most." (Dunbar & Rogers, *Principles Of Stratigraphy*, p 128, standard geology textbook used in universities).

21. Robert F. Helfinstine, and, Jerry D. Roth, *Texas Tracks and Artifacts*, 1994, p. 89-90. and a personal letter from Dr. Dale Peterson, M.D., March 27, 2002, stating: "I was first able to view what I now believe to be a fossilized human finger shortly after it was excavated. At that time my comment was 'interesting.' The fossil clearly had the shape of a human finger. It had a fine taper of the tip, typical of a female finger. Male fingers tend to be a bit more blunt. The fingernail and cuticle were clearly visible and perfectly formed and proportioned. Nevertheless, I withheld judgment as to its authenticity knowing that rocks such as limestone can assume nearly any shape when they flow into a hole before setting up. Several years later I was privileged to view the fossil again after it had been sectioned. At that time I observed that the fossil was not of uniform or random density and coloration. The internal appearance of the fossil was identical to what one sees when a human finger is sectioned. The skin margins and subcutaneous tissue were clearly delineated. The bone matrix was clearly defined, and features consistent with flexor and extensor tendons were present. CT scans of the fossil likewise revealed the anatomical features of a human finger, as noted above. It is my professional opinion that the fossil unearthed at Glen Rose, Texas, is, in fact, a petrified human finger and not an infill of a wormhole or similar artifact."

22. Joe Taylor, *Fossil Facts and Fantasies*, (Crosbyton, Tex.: Mt. Blanco Publishing Co. 1999), p. 75.

23. Don R. Patton, personal collection.

24. Charles Gould, *Mythical Monsters*, 1884, p. 201, reprinted 1995 by Studio Editions Ltd., London, England.

25. Herodotus, the text of Canon Rawlinson's translation, A.J. Grant, editor, two volumes (New York, N.Y.: Charles Scribner's Sons 1897). Volume 1, Book II, chap. 5.

26. Josephus, *Antiquities of the Jews*, Book II, chap. 10.

27. Herodotus, volume 1, Book II, para. 75. Herodotus then goes on to describe the ibis bird in paragraph 76. A footnote in the Rawlinson edition (p. 176, note 1) then points out that many specimens of this particular ibis bird (Numenis ibis) have been found mummified in Egypt. George Cuvier, in his *Researches sur les ossemens fossiles des quadrupeds*, 1812, remarked that these mummified specimens were the same as the modern bird and if in over 3,000 years there has been zero modification [evolution] one may multiply 3,000 by

zero and the modification would still be zero! He had a couple of nice engravings side by side showing the skeleton of the mummified bird and a modern-day ibis bird.

28. *Pliny's Natural History*, Book VIII, chap. 9, translated by J. Bostock and H.T. Riley; Bohn, (London, England: 1855), as in Gould's *Mythical Monsters*, p. 169.

29. Wm. Caxton 1484. Aesop. Folio 138, as quoted by Bill Cooper in *After the Flood* (West Sussex, England: New Wine Press, 1994), p. 139.

30. Theodore DeBry, book of engravings from America, ca. 1562.

31. Don R. Patton, primary research interviews with Dr. Cabrera. Don has worked with Dr. Dennis Swift in gathering information about these remarkably unique engravings.

32. Randall A. Reinstedt, *Shipwrecks and Sea Monsters of California's Central Coast* (Carmel, Calif.: Ghost Town Publications, 1975), p. 160.

33. John Goertzen, "New *Zuiyo Maru* Cryptid Observations: Strong Indications It Was a Marine Tetrapod," Creation Research Society Quarterly vol. 38 (June 2001), p. 19ff.

34. Richard Carrington, *Natural History*, 66:183-187, 1957, in *Incredible Life: A Handbook of Biological Mysteries*, William Corliss, The Sourcebook Project, p. 519.

35. Duane Gish, *Dinosaurs By Design*, (El Cajon, Calif.: Master Books, 1992), p. 16.

36. *National Geographic* (Dec. 1968).

37. "A Bolivian Saurian," *Scientific American*, 49:3, (1883), in *Incredible Life: A Handbook of Biological Mysteries*, William Corliss, p. 531.

38. Roy P. Mackal, *A Living Dinosaur? In Search of Mokele-Mbembe*, (Leiden, Netherlands: E.J. Brill, 1987).

39. *Journal of Paleontology*, vol. 61 no. 6, 1986-7, p. 198-200.

Section 4

Unlocking the Mysteries of Ancient Civilization

1. "A Lady from China's Past," *National Geographic*, (May 1974).

2. ibid.

3. Korium Megertchian, "Metallurgic Factory," *Bible-Science Newsletter*, Five Minutes section, Feb. 1973, p.3, and Rene Noorbergen, *Secrets of the Lost Races*, (New York, N.Y.: Barnes and Noble Books, 1977), p. 32.

4. "Ancient Electroplating," 1933 *Annual Log*, (New York, N.Y.: Scientific American Publishing Co., 1933), p. 85.

5. The word zodiac comes from the Greek word *zodiakos* meaning a circle, but derived from the primitive root *zoad* which denotes *A Way* or *A Path* or *Going by Steps*.

6. Joseph A. Seiss, *The Gospel in the Stars*, Castel Press, 1884, (Grand Rapids, Michigan: Kregel, 1979); Kenneth C. Fleming, *God's Voice in the Stars: Zodiac Signs and Bible Truth*, (Neptune, New Jersey: 1981); D. James Kennedy, *The Real Meaning of the Zodiac*, (Fort Lauderdale, Flor.: Coral Ridge Ministries, 1997).

7. Request further resources on the Flood from the publisher of this book. A key reference is *The Genesis Flood*, by H. Morris and J. Whitcomb, (Presbyterian and Reformed Publishing Co. 1961).

8. Noorbergen, Rene, *Secrets of the Lost Races*, (New York, N.Y.: Barnes and Noble Books, 1977), p. 202.

9. James George Frazer, *Folklore in the Old Testament*. 3 vols. (London, England: Macmillan & Co., 1918). In Volume 1, p. 104-360, Frazer provides 138 accounts of the Flood from separate cultures from around the world and he does document minutely. It is also beautifully written though the author did not accept the Bible as authoritative history or divine inspiration.

10. Ibid. p. 26.

11. As cited in *Ancient Man: A Handbook of Puzzling Artifacts*, by William Corliss, (The Sourcebook Project, 1978), p. 656.

12. G. Frederick Wright, *American Antiquarian*, 11:379-381 (1889), as in William Corliss, *Ancient Man: A Handbook of Puzzling Artifacts*, p. 458.

13. Robert Helfinstine, and Jerry Roth, *Texas Tracks and Artifacts*, (Anoka, MN, 1994), p. 91.

14. Wil Durant, *The Story of Civilization: Our Oriental Heritage* (New York, N.Y.: Simon & Schuster, 1942), p. 224, 225, as in Ethel Nelson's and C.H. Kang's *The Discovery of Genesis* (St. Louis, Mo.: Concordia Publishing House, 1979), p. 106.

15. Ethel R. Nelson, *God's Promise to the Chinese*, (Dunlap, Tenn.: Read Books, 1997), p. 2.

16. James Legge, *The Notions of the Chinese Concerning God and Spirits* (Hong Kong: Hong Kong Register Office, 1852), p. 28, as in Kang and Nelson's *Discovery of Genesis*, p. 15.

17. Ruth Beechick, *Genesis, Finding Our Roots* (Pollock Pines, Calif.: Arrow Press, 1997), p. 35.

18. Nelson, *God's Promise to the Chinese*, p. 109.

19. The characters contained in these pages are the **traditional** Chinese characters. A number of years ago, the Communist Chinese sought to revise the characters to a "**simplified**" form. They did this for two reasons: (1) to make typesetting easier; and (2) to eliminate from their written language the built-in references to God. I am greatly indebted in this research to C.H. Kang and Ethel R. Nelson – authors of *The Discovery of Genesis*. I would highly recommend your ordering and reading their book.

20. Charles Hapgood, *Maps of the Ancient Sea Kings*, (Philadelphia, Penn.: Chilton, 1966).

21. Barry Fell, *America B.C.*, (New York, N.Y.: Simon and Schuster, 1976), p. 21.

22. "Polynesians," *National Geographic*, (Dec. 1974).

23. *The World's Last Mysteries*, (Pleasantville, NY: Readers Digest Assoc., Inc., 1967), p. 94.

24. "The Mayan," *National Geographic*, v. 148:6, (Dec. 1975), p. 783.

25. Ibid., *The World's Last Mysteries*, p. 138.

26. Harry Schwalb, "Electric Batteries of 2,000 years ago," *Science Digest*, 41:17-19 (April 1957), in William Corliss' *Ancient Man...*, p. 453 and *Feats and Wisdom of the Ancients*, Time Life, p. 20.

27. Derek J. de Solla Price, "Unworldly Mechanics," *Natural History*, 71:8-17, (March 1962) in Wm. Corliss, *Ancient Man: A Handbook of Puzzling Artifacts*, The Sourcebook Project, 1978, p. 451, and *Feats and Wisdom of the Ancients*, Time Life, p. 93.

28. *Feats and Wisdom of the Ancients*, (Alexandria, Va.: Time-Life Books, 1990), p. 28.

Recommended Resources

So many excellent creation books have been published recently that it is difficult to highlight just a few. Many are out of print but some are now on the Internet. Serious students should take advantage of the many resources on the World Wide Web. A good place to start is with the links page of the Creation Resource Foundation at **www.creationresource.org**. There you will find links to excellent research and articles at many levels of study.

Among the highest recommendations, the excellent works of Dr. Henry Morris will benefit everyone. The founder of the Institute for Creation Research, Dr. Morris is a brilliant scientist and a devout biblical scholar. His writing is both profound and easily understood by even young readers. The following are some of his basic works considered essential to any beginning home library on origins.

Scientific Creationism by Dr. Henry Morris covers the many key scientific issues relevant to origins including astronomy, biology, fossils, earth's age, man's origin, the Flood, and dealing with attempts to compromise the clear meaning of scripture.

The Biblical Basis for Modern Science by Dr. Henry Morris is a rich treasure providing a thorough foundation for a sensible view of reality that is consistent with both conservative biblical authority and observations of true science. The helpful appendices at the end are an important resource for every serious student.

Many Infallible Proofs by Dr. Henry Morris is ideal for those wanting a clear, easily understood, and logical defense of the authority of Scripture. Proofs from history and science, as well as a complete explanation of the authenticity of the Old and New Testaments, and the facts of the life, death, and resurrection of Christ make this text essential reading for every Christian.

The following authors spent lifetimes researching and assembling what is amazingly hidden from the general public. The complete media blackout of this data only makes sense in view of the fact that there is a very real spiritual war going on behind the scenes. The enemy of the Creator is real. You will be extremely enlightened by each of these excellent works.

In The Minds of Men by Ian Taylor is a well-illustrated, thorough record, from the Greeks to the UN, of the growing deception of evolutionism. "How can so many smart people believe the fairy tale of evolution?" He reveals how one deception after another has confused generations of educated people.

Evolution: The Fossils Still Say No! by Dr. Duane Gish is important reading on the silent witness of fossils. While many have been misled to believe that fossils "prove" evolution, in reality, they actually prove the truth of a global Flood.

Dinosaurs by Design by Dr. Duane Gish gives parents the ideal answer to colorfully re-paint the picture that children usually hear about dinosaurs. Many dinosaur types are illustrated and described. Evidence links dinosaurs to the dragon stories of historic people. This puts them in a context of truthful world history that makes sense.

Tornado in a Junkyard by James Perloff covers the evolution debate in a very readable style. Jim was a rebellious atheistic intellectual who eventually was challenged to do his homework. In the process, his detective skills led him to well-documented information. The classic misconceptions of evolution, the fossil frauds, and how Hollywood has twisted science and history, make this THE BOOK to put in the hands of any skeptic you know.

Eternity In Their Hearts by Don Richardson is surprisingly eye opening. In a fascinating series of documented stories, from the Santal people in India to the educated philosophers of ancient Athens, Don tells how the concept of a supreme Creator has existed for centuries in many cultures of the world.

Video – ***A Question of Origins***, produced by Jim Tetlow, is essential viewing for every public school student you know. Using high tech digital images, orchestrated with brilliant quotes from respected scientists, this is a thorough one-hour presentation that leaves skeptics without excuse. It covers the big bang, the complexity of living cells, and the bankruptcy of animal and human evolution. Brilliant computer graphics and insightful commentary by many respected leaders make this one of the most viewed videos on origins.

Video set – ***The Wonders of God's Creation***, produced by Moody Video, is three hours of some of the most compelling evidences of God's design ever filmed. Covering Planet Earth, Animals, and Human Life, these programs are classic core curriculum for every student of nature and the Bible. Brilliant photography, expert scientific commentary, and background music to inspire you with the Creator's awesomeness puts this set at the top of your resource list.

Video tapes – by Creation Research of Australia with John Mackay include **some of the most informative and stimulating creation films made.** Television quality programs and lecture tapes are all packed with thought provoking detail on dinosaurs, ancient cultures, flood geology, living fossils and biology. Over 20 programs are available. Request a catalog from AWE, Box 570, El Dorado, CA 95623, USA.

Darwin's Enigma by Luther Sunderland is unique for revealing that chief executives of five major natural history museums of the world acknowledge that their exhibits are built on presuppositions without the evidence to support them.

The Gospel in the Stars by Joseph Seiss is one of the most eye-opening, in-depth studies of the true significance of the names of the stars and constellations. Published in 1882, this book is one of the most popular among students of science and the Bible.

A complete catalog of excellent resources is available on the Internet at Awesome Works Emporium at www.creationresource.org or by mail if you send first class postage for 8 ounces to AWE, Box 570, El Dorado, CA 95623, USA, or call (866) 225-5229.

Web Sites on Creation and Related Topics

There are so many websites! Where do you start? To discover the wide variety of creation organizations, start your browsing at the Creation Resource Foundation at

www.creationresource.org

You'll find educational tools, links and an on-line store. Your support and patronage is greatly appreciated to continue the work of educating others on vital foundational truths. To access articles and on-line books dealing with the origins issue, you will find a wealth of information at the following sites:

www.creationism.org is an Internet informational ministry devoted to providing a broad base of referral data taking you in many directions and in different languages.

www.evolution-facts.org has a massive collection of scientific facts against evolution. This is one of the most complete anti-evolution websites, containing a fabulous amount of scientific data disproving evolutionary theory to build your research papers.

www.parentcompany.com/csrc/ is from the Creation Science Research Center with books on-line like the *Handy Dandy Evolution Refuter* by Robert Kofahl.

www.rae.org is from the Revolution Against Evolution, a ministry of a church and the collecting efforts of Doug Sharp, and includes a diverse assortment of articles.

www.ldolphin.org is the on-line library of Lambert Dolphin, a research scientist who has gathered an extensive collection of articles on many biblical subjects including origins.

www.creationeducation.org provides useful articles and slide shows you can use to teach others.

www.creationmoments.com enables you to listen to daily Creation Moments radio brodcasts or print a copy of the transcripts. Inspiring brief articles on wonderful examples of design.

www.icr.org lists all the back issues of *Acts and Facts* published for many years by the highly esteemed ministry of the Institute for Creation Research.

www.christiananswers.net contains a wealth of creation articles plus indexes for other biblical subjects.

www.pathlights.com The "Creation-Evolution Encyclopedia" includes a step-by-step sequence of articles that discuss technical and biblical insights on a wide variety of origins topics.

www.apologeticspress.org lists topical articles very helpful to students wanting biblically based input on science issues.

www.raycomfort.com is a wonderful source of on-line and published information authenticating the supernatural origin of the Bible. Every Christian leader will appreciate this site.

Index

Unlocking the Mysteries of Creation

A Dynamic Multi-Media Seminar Exploring the
Wonders of Our World and Hidden Truths of the Bible

The very foundations of biblical faith are
being deliberately undermined,
and it's happening in the guise of what appears
to be science!

**Are the Bible accounts of creation, the Great Flood, and other
related issues really verified by honest investigation?**

This extraordinary seminar is a life-changing experience for the whole family!

It has been seen in hundreds of communities nationwide.

It is filled with colorful multi-media throughout the presentation.

This timely, unique, Bible-based presentation surprises viewers with colorful insights on:

- **Mysteries of Ancient Civilizations**
- **Dinosaurs and Cavemen**
- **Fossils and Evolution**
- **The Age of the Earth**

- **Mysteries of Life**
- **Extra-terrestrials**
- **Missing Links**
- **The Perfect Earth**

Available live and on video tape for your group, church or school.

"Excellent academically and scientifically!

…Excellent scripturally and theologically!

…Excellent in meekness and kindness!

You need to have this creation seminar in your church."

Pastor E. J. Gruen
Shawnee, Kansas

"…the case he [Dennis Petersen] made for creation was quite impressive… I am happy to recommend this seminar to others, believing that the case for creation is one that is vitally important in our age and that Christians must become better informed on the issue."

Dr. D. James Kennedy, Pastor
Coral Ridge Presbyterian Church
Fort Lauderdale, Florida

For further information on how your church or organization can schedule a seminar in your city, contact:

Creation Resource Foundation

P.O. Box 570, El Dorado, CA 95623 – (530) 626-4447 – mail@creationresource.org

www.creationresource.org

Dennis Petersen, B.S., M.A.

Born and raised in the San Gabriel Valley of Southern California, Dennis Petersen has pursued his lifetime interests in nature, history and photography both in school and professionally.

After earning his science degree and working as an exhibit preparator in historical museums, he received his M.A. from the State University of New York in museum administration. He spent several years as a museum curator and then had an encounter with God that changed his whole purpose in life.

In 1973, he was providentially led to attend a Canadian Bible college where he received three years of concentrated biblical training. For four years he taught apologetics and Old Testament survey at that Bible Institute, and also developed a college course on science and the Bible. He ministered as a church pastor for five years before the Lord sovereignly led him to begin the Mysteries of Creation seminar ministry in 1984.

Through seminars, publications, and media presentations, his keen desire is to realign this generation's thinking to the life-giving Word of God: the Only Key to unlock the ultimate mysteries.

Dennis founded the Creation Resource Foundation in El Dorado, California, where he and his lovely wife Viola have raised all four of their delightful children in the home they built in that rural foothill community. He has actively participated as a teaching elder in the local church there and has served many area churches and schools with specialized teachings on creation subjects. His friendly personality expresses the warmth of Christ in a genuine spirit of concern for people everywhere he goes. Mr. Petersen can be contacted for speaking engagements and conferences at the Creation Resource Foundation.

Picture Credits

Daniel Cordova – 26, 31, 33, 76, 77, 78, 79, 80, 81, 82, 87, 89, 91, 97, 99, 102, 103, 104, 105, 108, 112, 117, 118, 126, 128, 130, 131, 135, 139, 142, 163, 166, 167; **Michael McCartie** – 93, 95, 109, 125, 127, 151, 153, 169; **Kris Westbeld** – 103; **Jonathan Chong** – 140, 182, 183; **Warren Dayton** – 19, 42, 52, 83, 131, 162, 188; **NASA** – 9, 16, 23, 24, 27, 28, 30, 36, 46, 48, 50, 58, 59, 61, 74, 75, 173; **Bill Fraser** – 9, 125; **Roy Mackal** – 169; **Don Patton** – 144, 145, 155, 156, 157, 158, 159, 160, 161, 165; **Don Harris** – 148, 149; **Ian Taylor** – 164, 165, 167, 176; **Virgil Niggemeyer** – 116; **Clifford Wilson** – 188; **John Mackay** – 123 (Platypus photo Copyright 2001, Creation Research Australia, info@creationresearch.net); **Univ. California Santa Cruz** - 166; **Corel Professional Photos** – 53, 143, 176, 177, 181, 190, 193, 200, 201, 202, 203, 206, 207, 208, 214, 221; **Thomas Kinkade** – 2 and 3 (previously the cover of the earlier version of this book).

Information on pages 110-111 provided by **Bill Morgan** of the Orange County (California) Creation Science group. Reference his excellent web page at www.fishdontwalk.com .

(Attempts have been made to find sources to give credit for images on p. 204 and 210. Photographs not noted above were made by the author.)

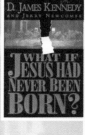 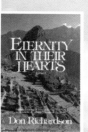